National Aeronautics and Space Administration

I0473763

Ikhana

Unmanned Aircraft System
Western States Fire Missions

Peter W. Merlin

National Aeronautics and Space Administration
NASA History Office
Washington, D.C.
2009

ii

Library of Congress Cataloging-in-Publication Data

Merlin, Peter W., 1964-
  Ikhana Unmanned Aircraft System : Western States fire missions / by Peter W. Merlin.
      p. cm.
  "August 2009."
  1. Aeronautics in forest fire control--West (U.S.) 2. Aerial photography in forestry--West (U.S.)
3. Drone aircraft. 4. Predator (Drone aircraft) 5. United States. National Aeronautics and Space
Administration. I. Title.
  SD421.43.M47 2009
  634.9'6180978--dc22

Contents

Acknowledgments

A great many people helped make this book possible. I am especially indebted to Tony Springer of the Aeronautics Research Mission Directorate at NASA Headquarters for sponsoring this project. Ikhana project manager Thomas K. Rigney, lead operations engineer Gregory P. Buoni, project engineer Kathleen Howell, and project pilots Mark Pestana and Herman Posada provided support and made numerous helpful comments. Vincent G. Ambrosia, senior research scientist, California State University, Monterey Bay; Dr. Steven S. Wegener, senior research scientist, Bay Area Environmental Research Institute; and Dr. Susan Schoenung, project scientist, NASA Ames Research Center, made sure my descriptions of the sensor systems and their development were accurate. Thanks to Sarah Merlin for copy editing of the manuscript. I am also grateful to Carla Thomas of the NASA Dryden Photo Branch for providing scanned images and to Steve Lighthill for layout and graphic design. Thanks also to Christian Gelzer for looking over the manuscript and for other technical assistance. Special thanks to the Ikhana team and to the firefighters who put the technological resources described in this monograph to good use combating wildfires in the western United States.

v

Preface

In 2006, NASA Dryden Flight Research Center, Edwards, Calif., obtained a civil version of the General Atomics MQ-9 unmanned aircraft system and modified it for research purposes. Proposed missions included support of Earth science research, development of advanced aeronautical technology, and improving the utility of unmanned aerial systems in general. The project team named the aircraft Ikhana – a Native American Choctaw word meaning intelligent, conscious, or aware – in order to best represent NASA research goals.

Researchers at Dryden have a long history of using remotely piloted research vehicles to expand the frontiers of knowledge. Among the first was the Hyper III, a Langley-designed lifting body. In support of the M2 lifting-body program of the early 1960s, R. Dale Reed had built a number of small lifting-body shapes and drop-tested them from a radio-controlled mothership. Reed and pilot Milton O. "Milt" Thompson wanted to try the remote flying concept on a full-scale design. The remotely piloted research vehicle, or RPRV, weighed 484 pounds, measured 32 feet in length, and spanned 18 feet. On December 12, 1969, the Hyper III was launched from a helicopter at 9,800 feet, glided three miles, reversed course and glided three miles more to the lakebed.

In 1975 a series of stall and spin tests was begun at the center with a group of 3/8-scale F-15 RPRVs. With it carried aloft by the center's B-52 mothership and released at about 50,000 feet, a pilot in a ground-based cockpit flew the RPRV (essentially a miniature F-15 that lacked engines) via instruments and a television monitor, stalling and recovering the aircraft to see what modifications worked best.

Flights of another aircraft, dubbed the "Mini-Sniffer," took place between 1975 and 1979, testing the concepts of an RPRV operating in the Martian atmosphere or conducting high-altitude atmospheric research around the globe. Again, the pilot remained in a ground cockpit to control the vehicle in flight.

The DAST – Drones For Aerodynamic and Structural Testing – program, a high-risk flight experiment using a ground-controlled, pilotless aircraft, was undertaken at Dryden from 1977 to 1983. Described by NASA engineers as a "wind tunnel in the sky," the DAST vehicle was a specially modified Teledyne-Ryan BQM-34E/F Firebee II supersonic target drone.

It was flown to validate theoretical predictions under actual flight conditions in a joint project with Langley Research Center.

From 1979 to 1983 the HiMAT (Highly Maneuverable Aircraft Technology) aircraft was flown, one of two subscale research vehicles meant to demonstrate advanced fighter technologies that have since been used in development of many modern high-performance military aircraft. About one-half the size of a standard manned fighter, and powered by a General Electric J85-21 jet engine, the HiMAT vehicles were launched from NASA's B-52 carrier aircraft at an altitude of about 45,000 feet. The aircraft were flown remotely by a NASA research pilot from a ground station, with the aid of a television camera mounted in the HiMAT cockpit that gave the pilot a forward field of view.

In 1984 Dryden moved from small-scale vehicles to full-size aircraft when a pilot intentionally crashed a retired Boeing jetliner onto Rogers Dry Lake to test a compound meant to reduce post-crash fires on airliners. With its fuel tanks filled with anti-misting kerosene (AMK), a pilot in a ground cockpit at Dryden flew the Boeing 720 from the Edwards Air Force Base main runway and, after gaining altitude, descended and crashed it onto the lakebed. To the dismay of the manufacturer as well as the Federal Aviation Administration, which was ready to issue a requirement calling for AMK to be used in all commercial aircraft, the airliner burst into a fireball and burned, eliminating the additive from potential use.

And Dryden was the center for operations of a family of solar-powered aircraft designed to explore the potential for such aircraft to monitor Earth's atmosphere as well as such other factors as moisture content in soil. Beginning in the 1990s, Pathfinder, Pathfinder-Plus, and Helios were all part of the Environmental Research Aircraft and Technology, or ERAST, program through which researchers hoped to mature RPRV and unmanned aerial system technologies.

Building on experience with these and other unmanned aircraft, NASA scientists developed plans to use the Ikhana for a series of missions to map wildfires in the western United States and supply the resulting data to firefighters in near-real time. A team at NASA Ames Research Center, Mountain View, Calif., developed a multispectral scanner that was key to the success of what became known as the Western States Fire Missions. Carried out by team members from NASA, the U.S. Department of Agriculture Forest Service, National Interagency Fire Center, National Oceanic and Atmospheric Administration, Federal Aviation Administration, and General Atomics Aeronautical Systems Inc., these flights represented an historic achievement in the field of unmanned aircraft technology.

Introduction

From the earliest days of aviation, pilots have enjoyed a certain amount of glamour because of their apparent willingness to face danger in the skies (military or stunt pilots) or their ability to travel to exotic locales (airline and cargo pilots). Long before the Wright brothers made their first powered flight in December 1903, there were also exciting developments in the field of unmanned aviation.

On May 6, 1896, Dr. Samuel Pierpont Langley's steam-powered, pilotless Aerodrome No. 5 successfully completed two flights, each lasting 1.5 minutes, covering a distance of about half a mile and reaching a maximum altitude of 100 feet. Just three months prior to the Wrights' historic December 1903 flight at Kitty Hawk, N.C., German designer Carl Jatho demonstrated a gasoline-fueled pilotless biplane that covered a distance of 196 feet at a height of 11 feet – further and higher than the Wright Flyer. During World War I and World War II, radio-controlled aircraft served as aerial torpedoes and aerial targets, roles that evolved into the development of cruise missiles, airborne decoys, and target drones. In the late 1940s and 1950s, drone aircraft were pressed into service for such tasks as flying through clouds of radioactive fallout from nuclear explosions to collect samples without endangering aircrews. In the 1960s, improvements in unmanned vehicle technology spawned development of tacti-

In May 1896, Dr. Samuel Pierpont Langley's steam-powered Aerodrome No. 5 made the world's first successful flight of an unpiloted, engine-driven, heavier-than-air craft. It was launched from a spring-actuated catapult mounted on top of a houseboat on the Potomac River near Quantico, Virginia. The airplane is displayed at the Smithsonian Institution's National Air and Space Museum in Washington, D.C.
Courtesy of Brian Nicklas

The General Motors A-1 flying bomb was a conventional high-wing monoplane powered by a 200-hp piston engine. It was launched from a four-wheeled dolly, and could carry a 500-lb bomb over a distance of 400 miles. Initial tests between 1941 and 1943 were not very successful because basic problems of inadequate control had not been resolved.
AFFTC History Office

cal and strategic reconnaissance platforms. By the dawn of the 21st century, unmanned aircraft systems (UAS) – i.e., airplanes without pilots on board – were used more and more frequently for a variety of missions.[1]

The term "unmanned aircraft system" was adopted by the Department of Defense (DoD) and Federal Aviation Administration (FAA) to replace the term "unmanned aerial vehicle" (UAV) to better represent the fact that more is involved than hardware that flies though the sky. A UAS consists of an aircraft, ground-based control system, data link, and related support equipment.

While unmanned systems will never entirely replace aircraft with onboard crews, they are useful in performing so-called dull, dirty, or dangerous missions. Because a UAS can be autonomous or remotely piloted, aircrew members are not subjected to the risks ordinarily associated with long-duration or high-altitude flight and the hazards inherent in performing certain tasks, such as flying through smoke, turbulence, airborne radioactive particles, or chemical residue.

[1]Hugh McDaid and David Oliver, *Smart Weapons: Top Secret History of Remote Controlled Airborne Weapons*, Orion Media, London, England, 1997.

On a long-duration flight, for example, the length of time an airplane can remain airborne is determined by such factors as crew fatigue and use of consumables (food, oxygen, etc.). With a remotely operated UAS, several pilots can alternate shifts in the ground control station (GCS) as often as necessary. Flight duration is limited only by fuel consumption and mechanical reliability.

Use of unmanned aircraft prevents aircrew exposure to such hazardous conditions as extreme weather, radiation, or in the case of military UASs, hostile fire. This eliminates the risk of physical harm to the aircrew and reduces political costs in the wake of mission failure. These characteristics have motivated military and civilian agencies to expand the use of UASs wherever feasible.[2]

gation systems, among others. During the first 50 years of powered flight, efforts – primarily involving military requirements – focused on guidance and navigation, stabilization, and remote control. In the half-century that followed, designers worked to improve technologies to support these capabilities through integration of improved avionics, microprocessors, and computers.[3]

The Airborne Antarctic Ozone Experiment (AAOE), an international, multi-institutional effort to study a sudden and unanticipated decrease in ozone over Antarctica, was a milestone. In July 1989 a workshop was held in Truckee, Calif., in conjunction with a meeting of the AAOE science team to address the science community's UAS needs. Known as the "Truckee Report," the resultant white paper

Perseus-A, a remotely piloted, high-altitude research vehicle designed by Aurora Flight Sciences Corp., was towed aloft by a ground vehicle and its engine started after it became airborne. The airplane reached an altitude of 50,000 feet on its third test flight.
NASA

The cost and complexity of robotic and remotely piloted vehicles is generally less than those of comparable aircraft that require an onboard crew because there is no need for life-support systems, escape and survival equipment, or hygiene facilities. UAS development has also contributed significantly to many technological innovations in aviation. Examples include autopilot systems, data links, and inertial navi-

detailed the need for unmanned science aircraft and the goal of making measurements from an altitude of 85,000 feet.

In the early 1990s, NASA's Earth Science directorate received a solicitation for research to support the Atmospheric Effects of Aviation Project, specifically to assess the potential impact of a commercial supersonic transport aircraft. Here again measurements were needed at 85,000 feet. Aurora Flight

[2] Office of the Secretary of Defense, "Unmanned Aircraft Systems Roadmap 2005-2030," http://www.acq.osd.mil/usd/Roadmap%20Final2.pdf, 2005, accessed Oct. 12, 2008.

[3] Ibid.

As with the Perseus-A, the Perseus-B was flown remotely by a pilot from a mobile flight control station on the ground. A Global Positioning System unit provided navigation data for continuous and precise location during flight. The ground control station featured dual independent consoles for aircraft control and systems monitoring.
NASA

Sciences of Manassas, Va., proposed developing the Perseus-A and Perseus-B remotely piloted research aircraft as part of NASA's Small High-Altitude Science Aircraft (SHASA) program.

In 1993, the SHASA effort expanded as NASA teamed with industry partners for the Environmental Research and Sensor Technology (ERAST) project. The goal of the ERAST effort was to develop and demonstrate aeronautical technologies applicable to remotely or autonomously operated aircraft capable of carrying out long-duration airborne science missions. Initial ERAST program demonstrations transferred technology to an emerging UAS industry and validated the capability of unmanned aircraft to carry out operational science missions.

The program was managed at NASA's Dryden Flight Research Center at Edwards, Calif., with significant contributions from the agency's Ames, Langley, and Glenn research centers. Industry partners included such aircraft manufacturers as AeroVironment, Aurora Flight Sciences, General Atomics Aeronautical Systems Inc., and Scaled Composites. Sensors to be carried by the research aircraft were developed by Thermo-Mechanical Systems, Hyper-

spectral Sciences, and Longitude 122 West.[4]

General Atomics developed a civilian model of an upgraded version of the company's Predator military reconnaissance platform – called Predator B – to meet specific requirements of NASA's Earth Science Enterprise for a flight-validated, consumable-fuel UAS to perform on-location science missions. Known as Altair®, it exceeded minimum performance requirements and set the stage for the Ikhana UAS.

[4]"ERAST: Environmental Research and Sensor Technology Fact Sheet," NASA Dryden Flight Research Center, Edwards, CA, 2002.

x

Chapter One
DON'T FEAR THE REAPER

The Ikhana's heritage is traced to a family of medium-altitude, long-endurance UAS vehicles developed by General Atomics Aeronautical Systems Inc. (GA-ASI) of San Diego, Calif., in the late 1980s and early 1990s. The first of these was the Gnat 750, a long-endurance tactical surveillance and support system designed for large payload capacity, ease of use, and low operating cost. Following its first flight in 1989, the Gnat 750 demonstrated un-refueled flight duration of up to 30 hours, 25,000-foot service ceiling, and climb rate of 1,100 feet per minute. With a length

The Central Intelligence Agency has operated a Gnat 750-45, called Lofty View, with expanded capabilities. It is reportedly capable of carrying a 450-500-pound payload consisting of synthetic-aperture radar (SAR) with 12-inch resolution, three electro-optical (EO) or infrared (IR) sensors in a chin turret, and a wideband satellite data-link antenna to allow transmission of real-time data in the form of video.

The most recent improved version, called I-Gnat, has been reconfigured with a turbo-charged engine to increase its operating altitude to 30,500 feet for up to 48 hours.[5]

In 1997 NASA researchers used a General Atomics Gnat 750, under the name Altus, to demonstrate the ability to cruise at altitudes above 40,000 feet for sustained durations on atmospheric science missions.
NASA

of 16 feet and a 35-foot wingspan, the vehicle has a gross takeoff weight of 1,140 pounds including a 330-pound payload.

The Gnat 750 was extensively field-tested during deployments to Albania, Bosnia, and Croatia to monitor military facilities, battlefield entrenchments, supply-caches, and troop movements. The need to relay data from the UAS through a manned aircraft that could remain on station for only about two hours at a time greatly limited the overall effectiveness of the system.

Gnat Grows Up

The RQ-1A/MQ-1A Predator (sometimes called Predator A) was introduced in 1994 as a growth version of the Gnat 750. Originally serving as an Advanced Concept Technology Demonstrator (ACTD) for the Air Force, the Predator soon became the most com-

[5] Maj. William G. Chapman, *Organizational Concepts for the Sensor-to-Shooter World – The Impact of Real-Time Information on Airpower Targeting*, Air University Press, Maxwell Air Force Base, AL, May 1997.

The MQ-1A Predator of the 11th Reconnaissance Squadron flies over the Nevada Test and Training Range near Creech Air Force Base. The Predator provides armed reconnaissance and battlefield support to ground troops.
USAF/MSgt. Scott Reed

bat-proven UAS in the world, providing continuous and persistent armed reconnaissance and battlefield support to ground troops. The Predator ACTD was used in various training exercises, demonstrations, and operational deployments before graduating to fully operational status in 1996 with deployments to Southwest Asia and Europe.

The Predator is a medium-altitude endurance UAS with a length of 27 feet and a 48.7-foot wingspan. It has an endurance of 40 hours with 24 hours on station at a range of 500 nautical miles (nmi), and a service ceiling of 26,000 feet. The vehicle has an empty weight of 1,130 pounds and a maximum gross takeoff weight (GTOW) of 2,250 pounds. It is equipped with a C-band line-of-sight communications system, as well as UHF and Ku-band satellite data-link capabilities. The Predator is typically equipped with EO sensors to provide color video and IR sensors for night-vision capability. It can also carry a SAR payload plus laser designation, spotting, and range-finding systems. An

armed version, called the MQ-1, carries air-to-air or air-to-ground missiles.[6]

An Unmanned Aircraft System for NASA

In 1996, General Atomics produced two civil variants of the Predator called Altus (Latin for "high"). Similar in appearance to the Gnat 750, Altus was 23.6 feet long and featured long, narrow, high-aspect-ratio wings spanning 55.3 feet. Powered by a rear-mounted Rotax 912 turbocharged four-cylinder piston engine rated at 100 hp, the vehicle was capable of cruising at 80 to 115 mph. An 84-inch-diameter, two-blade propeller was used for flight up to 53,000 feet altitude. A 100-inch lightweight carbon-fiber propeller was available for flights at higher altitudes. Altus had a

[6] Dr. Maziar Arjomandi, "Classification of Unmanned Aerial Vehicles," course material for Mechanical Engineering 3016, University of Adelaide, Australia, 2007.

330-pound payload capacity in a nose compartment designed to accommodate a variety of sensors and scientific instruments.

NASA Dryden personnel initially operated the Altus vehicles as part of the ERAST program. Altus II, the first of the two aircraft to be completed, made its first flight on May 1, 1996. During developmental tests that summer, it reached an altitude of 37,000 feet. A few months later, researchers flew the Altus II in an Atmospheric Radiation Measurement-Unmanned Aerospace Vehicle (ARM-UAV) study sponsored by the Department of Energy's Sandia National Laboratories. During the course of the project, Altus II set a single-flight endurance record for remotely operated aircraft by remaining aloft for 26.18 hours through a complete day-to-night-to-day cycle.[7]

The ARM-UAV program was designed for col-

pabilities available among government agencies, universities, and private industry. Sandia provided overall technical direction, logistical planning and support, data analysis, and a multi-spectral imaging instrument. NASA's Goddard Space Flight Center and Ames Research Center, Lawrence Livermore National Laboratory, Brookhaven National Laboratory, Colorado State University, and the University of California Scripps Institute provided additional instrumentation. Participating university scientists were also drawn from the University of Maryland, the University of California at Santa Barbara, Pennsylvania State University, the State University of New York, and others.[8]

In September 2001, Altus II carried a thermal imaging system for the First Response Experiment (FiRE) during a demonstration at the General Atomics flight operations facility at El Mirage, Calif. A sensor

Altus II, seen here over Gray Butte, Calif., was flown in an atmospheric-radiation-measurement study. During a 26-hour mission, the Altus II set a single-flight endurance record for remotely operated aircraft.
NASA

lecting data on radiation/cloud interactions in Earth's atmosphere in order to better predict temperature rise resulting from increased carbon dioxide levels. Sandia's payload consisted of state-of-the-art radiometric instruments and devices to measure temperature, pressure, and concentration of water vapor.

The multi-agency program brought together ca-

developed for the ERAST program and previously used to collect images of coffee plantations in Hawaii was modified to provide real-time, calibrated, geolocated, multi-spectral thermal imagery of fire events. INMARSAT ground-data terminals, with antennas that adhered to the side of the Altus payload fairing, gave the Altus vehicle satellite communication. In

[7] "ALTUS II – How High is High?" NASA Fact Sheet FS-1998-12-058 DFRC, NASA Dryden Flight Research Center, Edwards, CA, 1998.

[8] W. R. Bolton, "Measurements of Radiation in the Atmosphere," NASA Tech Briefs, DRC-98-32, http://www.techbriefs.com/Briefs/Sep98/DRC9832.html, 1998, accessed June 10, 2009.

The MQ-9 Reaper, also known as Predator B, has greater payload and performance capabilities than the MQ-1. This made it an ideal candidate for NASA research missions.
USAF/MSgt. Robert W. Valencia

order to establish communication, operators steered the UAV on a bearing that pointed the antenna toward the satellite. This scientific demonstration showcased the capability of a UAS to collect remote sensing data over fires and relay the information to fire management personnel on the ground.[9]

Reaper Development

In 1999, General Atomics officials initiated development of a larger, turboprop-powered UAS called Predator B, later designated MQ-9 Reaper. While the company sought to expand the Predator's reliability and performance capabilities to meet ever-increasing mission requirements for civil and military applications, NASA scientists were interested in developing an aircraft with larger payload capacity, high-altitude

performance, and long endurance for Earth science missions.

Development of Predator B became a jointly funded effort in partnership with NASA as part of the ERAST program in January 2000 after the agency selected Altair, one of several competing proposals for development that met the agency's Earth Science Enterprise (now the Earth Science Division of NASA's Science Mission Directorate) UAS requirements. As joint partners in the project, which included flight validation as well as development of the aircraft, NASA's Office of Aerospace Technology invested approximately $10 million, while General Atomics contributed additional funds, with about $8 million earmarked for the Altair project.

Design criteria included the basic reliability of the Predator airframe, avionics, mechanical systems, data link, and flight control technology. Payload capacity was increased 500 percent (compared to the original Predator model) to accommodate improved EO/IR sensors, SAR, targeting radar, and up to 3,000 pounds

[9] Vincent Ambrosia, "Remotely Piloted Vehicles as Fire Imaging Platforms: The Future is Here!" NASA Ames Research Center, Moffett Field, CA, 2002.

The Altair operated by NASA had an 86-foot wingspan to optimize the aircraft for high-altitude flight. The Altair was designed to carry a 3,000-pound payload to 52,000 feet.
NASA

of externally mounted ordnance for its military role. Designers sought an endurance of over 30 hours, speeds greater than 240 knots true airspeed, and operational altitudes above 50,000 feet. A fault-tolerant, redundant flight-control system, and triplex avionics increased reliability and safety.[10]

Development of the Reaper began with Predator B-001, a proof-of-concept prototype. It had a larger fuselage than the standard Predator airframe with a wingspan of 66 feet. It was distinguishable from the original Predator by its Y-shaped tail and ventral vertical fin. A 950 standard horsepower (shp) Garrett AiResearch TPE-331-10T turboprop engine, de-rated to 700 shp, drove a rear-mounted, three-blade, controllable-pitch propeller, giving the aircraft a maximum speed of 220 knots. The B-001 had a 25-hour endurance and was capable of carrying a 750-pound payload to an altitude of 45,000 feet. B-001 had a maximum gross takeoff weight of 7,500 pounds.[11]

The Predator B-001 logged its first flight on Feb. 2, 2001, from the El Mirage facility. After completing initial airworthiness test flights and various software

and systems upgrades, a second series of test flights was flown in mid-summer 2001 to expand the flight envelope and validate autonomous flight capabilities. During a flight over the Edwards Air Force Base test range, the prototype reached a maximum sustainable altitude of 48,300 feet.[12]

While refining the configuration, General Atomics designers explored the possibility of building a jet-powered variant of the Predator B. It was initially proposed that a Predator B-002 with a 2,300-pound-thrust Williams FJ44-2A turbofan engine be built. NASA initially expressed some interest in a production version of the turbofan-powered variant, but development of such a vehicle was delayed until 2005 and a prototype, called Predator C Avenger, didn't fly until April 2007.

The Predator B-002 was instead built as a turbo-

[10] "General Atomics Fact Sheet," General Atomics Aeronautical Systems Company, San Diego, CA, 2007.

[11] "Altair / Predator B – An Earth Science Aircraft for the 21st Century," NASA Fact Sheet FS-073, NASA Dryden Flight Research Center, Edwards, CA, 2001. Additional information from David A. Fulghum and Bill Sweetman, "Predator C Avenger Makes First Flights," *Aviation Week & Space Technology*, Apr. 17, 2009.

[12] Ibid

prop-powered craft. It was nearly identical to B-001 but was equipped with single-string avionics as in the earlier Predator A model. The limitations of the B-002 design were dramatically highlighted by the capabilities of the improved Altair and Predator B-003 airframes, the first to incorporate triple-redundant avionics.

Altair was built with the 7,500-lb. GTOW fuselage and wing extensions. The Predator B-003 sported an upgraded 10,500-lb. GTOW airframe with an 86-foot wingspan. It was designed to carry a 3,000-pound

its ability to meet those standards, General Atomics conducted a multi-flight demonstration of the aircraft representative of a scientific data-gathering mission, including all integrated logistical support necessary when operating from a remote location. Demonstration plans included three long-duration, high-altitude flights with a payload consisting of imaging sensors and atmospheric-sampling instruments.

In the initial planning phase of the project, NASA scientists established a stringent set of requirements for Altair. These included mission endurance of 24 to

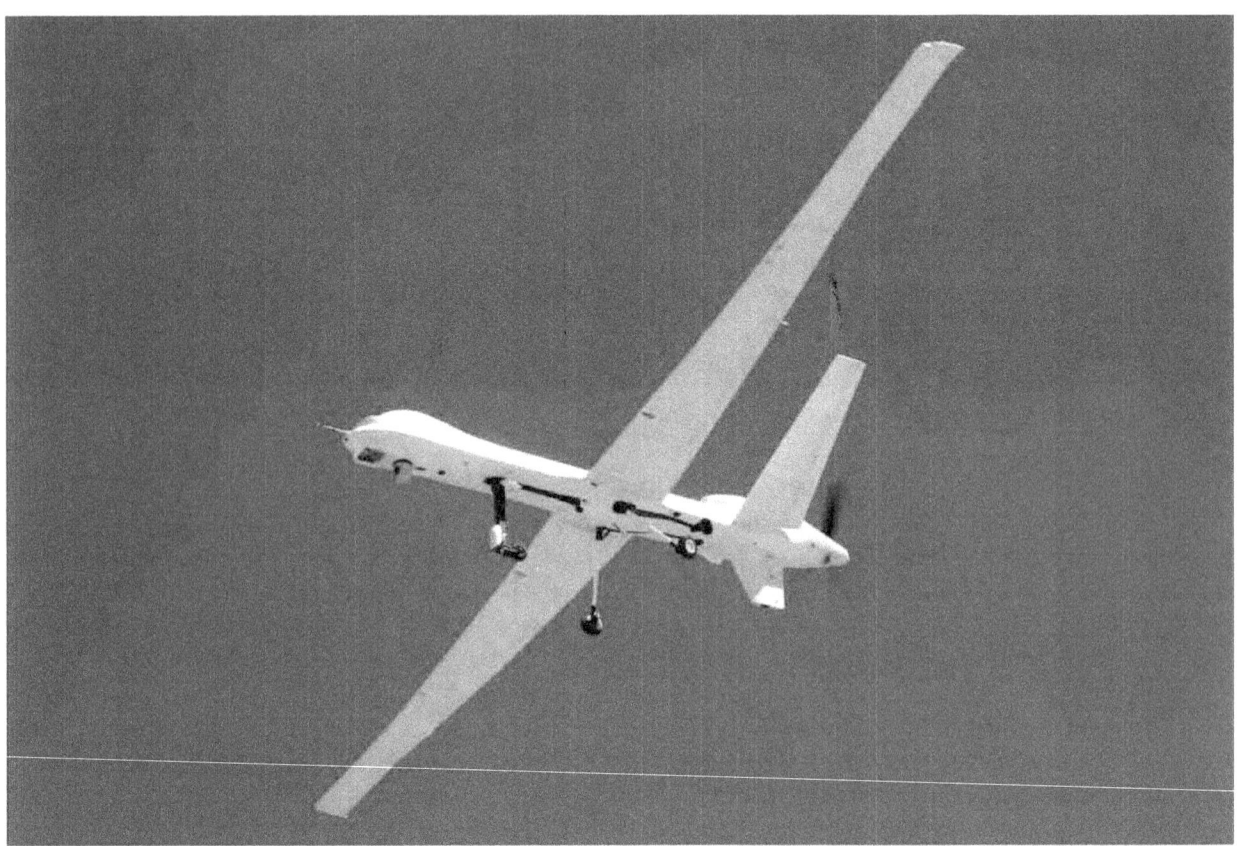

The Altair vehicle was the first UAV to feature triple-redundant avionics as well as a fault-tolerant flight control system. NASA

payload to 52,000 feet with a maximum endurance of more than 20 hours. After successful flight demonstrations at El Mirage NASA decided to lease Altair for use as a research platform.[13]

Altair

The Altair vehicle was designed to perform a variety of ERAST science missions specified by NASA's Earth Science Enterprise. To demonstrate

48 hours at an altitude range of 40,000 to 65,000 feet with a payload of at least 660 pounds. The project team also sought to develop procedures to allow operations from conventional airports without conflict with piloted aircraft. Additionally, Altair had to demonstrate command and control beyond-line-of-sight (BLOS) communications via satellite link, see-and-avoid operations relative to other air traffic, and the capability to communicate with FAA air-traffic controllers. In order to accomplish this, it would be necessary to equip the Altair with an automated collision-avoidance system and a voice relay to allow air-traffic control-

[13] Ibid.

lers to talk to ground-based Altair pilots. Since it was expected to be the first UAS to meet FAA requirements for operating from conventional airports, with piloted aircraft, in the national airspace (NAS), the aircraft also had to meet all FAA airworthiness and maintenance standards.[14]

The final Altair configuration was designed to fly continuously for up to 32 hours. Driven by its 700shp turboprop, it was capable of reaching an altitude of approximately 52,000 feet and had a maximum range of about 4,200 miles. It was designed to carry up to 750 pounds of sensors, radar, communications and imaging equipment in its forward fuselage.

Altair's first checkout flight at El Mirage on June 9, 2003, was a significant milestone in development of high-altitude, long-endurance, remotely operated aircraft. The aircraft lifted off gracefully and remained at relatively low altitude in the local area while the ground-based pilot evaluated its basic airworthiness and flight controls. In a post-flight briefing, NASA and GA-ASI officials were enthusiastic about the results.

"This is what we've been waiting for," said Glenn Hamilton, Altair project manager at NASA Dryden. "Now we can move forward with getting UAVs into the national airspace and conducting research."

Thomas J. Cassidy Jr., General Atomics Aeronautical Systems president and chief executive officer, echoed Hamilton's comments, saying, "Altair's first flight is a culmination of 10 years of experience in building reliable unmanned aircraft based on a common design philosophy."[15]

The Altair was the first UAV to feature triple-redundant controls and avionics for increased reliability, as well as a fault-tolerant, dual-architecture flight control system. After initial airworthiness test flights it served as an avionics testbed for the production version of the MQ-9 before being serving as a NASA research platform.

In July 2004, the Altair was deployed to Alaska to monitor fishing activities in the Bering Sea and the North Pacific Ocean for the U.S. Coast Guard but ended up being used to help map wildfires. Personnel stationed at the Poker Flat Research Range, 30 miles north of Fairbanks, helped secure crucial real-time imagery of wildfires raging through remote wilderness areas, including those in their own backyard.

Operators collected real-time information on fire locations, fire movement, and previously unidentified hot spots with the aircraft during a July 9 mission. Images of the so-called Boundary fire were used by firefighters as they tackled flames that charred portions of the 5,132-acre range in three separate waves that caused minor damage.[16]

In February 2004, General Atomics announced the award of a contract with the Canadian Forces to deploy the Altair in support of the Atlantic Littoral Intelligence, Surveillance, Reconnaissance Experiment – ALIX – in which the UAS was integrated with multi-mode maritime radar as well as electro-optical and infrared cameras for littoral and maritime surveillance off Canada's east coast. The month-long deployment commenced in August 2004 and involved BLOS operation of the aircraft and distribution of radar and video imagery to various end users throughout Canadian land, air, and maritime military forces.

Launched from an airfield at Goose Bay, Newfoundland, and flown on an instrument flight plan, control of the aircraft and payload was passed to a remote operations center in Ottowa for BLOS satellite communication operations. Surface-search radar and video imagery were passed to two remote video terminals (RVT) at separate locations.[17]

After returning to the U.S., the Altair was finally available for NASA research missions. The first task for Dryden researchers was to evaluate various new control, communications, and collision-avoidance technologies critical to enabling unmanned vehicles to fly safely in the national airspace. Three test flights for aircraft and payload evaluation were conducted at Gray Butte in April 2005. These flights included a full payload and reached a maximum altitude of 44,619 feet and 4.8-hour duration. NASA researchers were now ready to deploy the Altair for the first time.[18]

In May 2005, the National Oceanic and Atmo-

[14] Ibid.

[15] Alan Brown, "NASA's Newest Unmanned Aircraft Makes Successful First Flight," Press Release 03-193, NASA Headquarters, Washington, D.C., 2003.

[16] "Rocket launch secures photos for firefighting efforts," Alaska Science Outreach, http://www.alaskascienceoutreach.com/index.php/main_pages/catchitem/rocket_launch_secures_photos_for_firefighting_efforts/, accessed 10 June 2004.

[17] David W. Fahey, James H. Churnside, James W. Elkins, Albin, J. Gasiewski, Karen H. Rosenlof, Sara Summers, Michael Aslaksen, Todd A. Jacobs, Jon D. Sellars, Christopher D. Jennison, Lawrence C. Freudinger, and Michael Cooper, "Altair Unmanned Aircraft System Achieves Demonstration Goals," EOS Trans., No. 80. pp. 197-201, American Geophysical Union, Washington, D.C., 16 May 2006.

[18] "ALTAIR Unmanned Aircraft to Deploy to Canada," General Atomics Aeronautical Systems Company, San Diego, CA, 2004.

A Ku-band satellite communications system provided the Altair with uplink/downlink capabilities for beyond-line-of-sight control by a pilot in a ground station.
NASA

spheric Administration (NOAA) funded the UAV Flight Demonstration Project in cooperation with NASA and General Atomics. The experiment included a series of atmospheric and oceanic research flights off the California coastline to collect data on weather and ocean conditions, as well as climate and ecosystem monitoring and management.[19]

Science flights began on May 7 with a 6.5-hour flight to the Channel Islands Marine Sanctuary west of Los Angeles, a site thought ideal for exploring NOAA's operational objectives with a digital camera system and electro-optical/infrared sensors. The Altair carried a payload of instruments for measuring ocean color, atmospheric composition and temperature, and for surface imaging during flights at altitudes of up to 45,000 feet. Objectives of the experiment included evaluation of an unmanned aircraft system for future

scientific and operational requirements related to NOAA's oceanic and atmospheric research, climate research, marine sanctuary mapping and enforcement, nautical charting, and fisheries assessment and enforcement.[20]

Over the next few weeks, the Altair team made two attempts to conduct a 20-hour flight over the Pacific Ocean. Problems with the satellite communications link, however, resulted in flight durations of less than seven hours each. After a hiatus, flights resumed on Nov. 14 with a mission that lasted 18.4 hours and included ascent and descent altitude profiles at two fixed locations, a key aspect of the demonstration. The aircraft returned to Gray Butte somewhat earlier than planned on Nov. 15 due to fuel-management concerns. After landing, however, technicians found that fuel reserves would have permitted several hours

[19] Beth Hagenauer, "NOAA and NASA Begin California UAV Flight Experiment," Press Release 05-20, NASA Dryden Flight Research Center, Edwards, CA, 2005.

[20] Ibid.

The Altair was flown over California's Channel Islands in a joint NASA-NOAA project to collect data on atmospheric chemistry and the marine environment.
NASA

of additional operation.[21]

The joint NASA-NOAA research missions concluded on Nov. 16, 2005. During another flight over the Channel Islands, sensors on the aircraft gathered ocean color and atmospheric chemistry measurements and observed marine mammals and their environment. Researchers also evaluated capabilities that would be useful for conducting low-tide coastal mapping and NOAA law enforcement surveillance of the Channel Islands National Marine Sanctuary. NASA coordinated use of Altair with GA-ASI and provided mission management expertise to NOAA as well as flight planning.[22]

During these missions, the Altair flew in both restricted and controlled areas of the national airspace. The FAA was very cooperative with regard to flight plan approval and in-flight coordination with Altair through regional air traffic control centers along the West coast. The FAA had granted an Experimental Airworthiness Certificate for the Altair – the first ever for a UAS – in August 2005. The certification was an important step toward increasing the aircraft's freedom

to operate in the national airspace and recognized the quality and reliability of Altair operations. More than a dozen Experimental certificates have since been issued for unmanned vehicles, encouraging expanded development and use of UAS technology in U.S. airspace.[23]

In 2006, personnel from NASA, NOAA, GA-ASI and the U.S. Forest Service teamed up for the Altair Western States Fire Mission. The experiment demonstrated the combined use of a NASA Ames-designed thermal multi-spectral scanner integrated on a large payload capacity UAV, a data link telemetry system, near-real-time image geo-rectification, and rapid Internet data dissemination to fire center and disaster managers. The sensor system was capable of automatically identifying burned areas as well as active fires so there was no need to train sensor operators to analyze imagery. The success of this project set the stage for NASA's acquisition of an advanced Reaper variant and paved the way for future operational UAS missions in the national airspace.[24]

[21] Fahey, et al, "Altair Unmanned Aircraft System Achieves Demonstration Goals."

[22] Beth Hagenauer, "Altair UAV Flies Lengthy Science Missions For NOAA," Photo Release 05-73P, NASA Dryden Flight Research Center, Edwards, CA, 2005.

[23] Fahey, et al, "Altair Unmanned Aircraft System Achieves Demonstration Goals."

[24] "Altair Western States Fire Mission," http://ntrs.nasa.gov/archive/nasa/casi.ntrs.nasa.gov/20070031044_2007032019.pdf, accessed June 10, 2009.

Ground crewmen prepare the Ikhana for flight. The vehicle is a civilian version of the MQ-9, modified and instrumented for use in a variety of civil research roles.
NASA

Ikhana

In November 2006, NASA Dryden obtained a civilian version of the MQ-9 that was subsequently modified and instrumented for use in multiple civil research roles. These include supporting Earth science missions, development of advanced aeronautical technology, and acting as a testbed to develop capabilities for improving the utility of unmanned aerial systems.

The project team named the aircraft Ikhana, a Native American Choctaw word meaning intelligent, conscious or aware. The choice was considered descriptive of research goals NASA had established for the aircraft and its related systems at the time.[25]

"The name perfectly matches the goals we have for the aircraft," said Brent Cobleigh, NASA Dryden's first project manager for Ikhana. "They include collecting data that allow scientists to better understand and model our environmental conditions and climate, increasing the intelligence of unmanned aircraft to perform advanced missions, and demonstrating technologies that enable new manned and unmanned aircraft capabilities."[26]

The Ikhana is 36 feet long with a 66-foot wingspan. It can carry more than 400 pounds of sensors internally and over 2,000 pounds in external pods. The aircraft is

[25] "Ikhana Unmanned Science and Research Aircraft System," NASA Fact Sheet FS-097, NASA Dryden Flight Research Center, Edwards, CA, 2007.

[26] Beth Hagenauer, "Ikhana UAV Gives NASA New Science and Technology Capabilities," Press Release 07-12, NASA Dryden Flight Research Center, Edwards, CA, 2007.

"Ikhana" is a Native American Choctaw word meaning intelligent, conscious, or aware. The choice stemmed from NASA's research goals for the aircraft.
NASA

powered by a Honeywell TPE 331-10T turbine engine and is capable of reaching altitudes above 40,000 feet but with limited endurance at such altitudes. It is also the first production Predator B equipped with a digital electronic engine controller developed by Honeywell and General Atomics. This feature makes the Ikhana 5% to 10% more fuel efficient in some flight regimes than earlier versions of the aircraft.[27]

NASA's Science Mission Directorate was the first primary customer, employing the aircraft for Earth science studies. The Airborne Science Program uses both manned and unmanned aircraft to collect data within the Earth's atmosphere, complementing measurements of the same phenomenon taken from space and those taken on the Earth's surface. Ikhana is a versatile platform for such research because a variety of atmospheric and remote sensing instruments, including duplicates of sensors carried on orbiting satellites, can be installed to collect data during flights

lasting more than 20 hours.

"The need to collect data over day-night time cycles and over long distances in remote areas drives the need for a long-duration unmanned aircraft," said Cobleigh. "Piloted aircraft are limited by crew duty requirements that generally restrict science flights to 10 hours or less. Unmanned aircraft are also more suitable for remote missions spanning open oceans or the polar regions, where the lack of nearby emergency landing locations increases the risk for piloted missions."[28]

NASA's Aeronautics Research Mission Directorate also uses the aircraft for advanced systems research and technology development. Initial experiments included the use of fiber optics for wing shape and temperature sensing as well as control and structural loads measurements. Six hair-like fibers located on the upper surfaces of the Ikhana's wings

[27] "Ikhana Unmanned Science and Research Aircraft System."

[28] Beth Hagenauer, "Ikhana UAV Gives NASA New Science and Technology Capabilities."

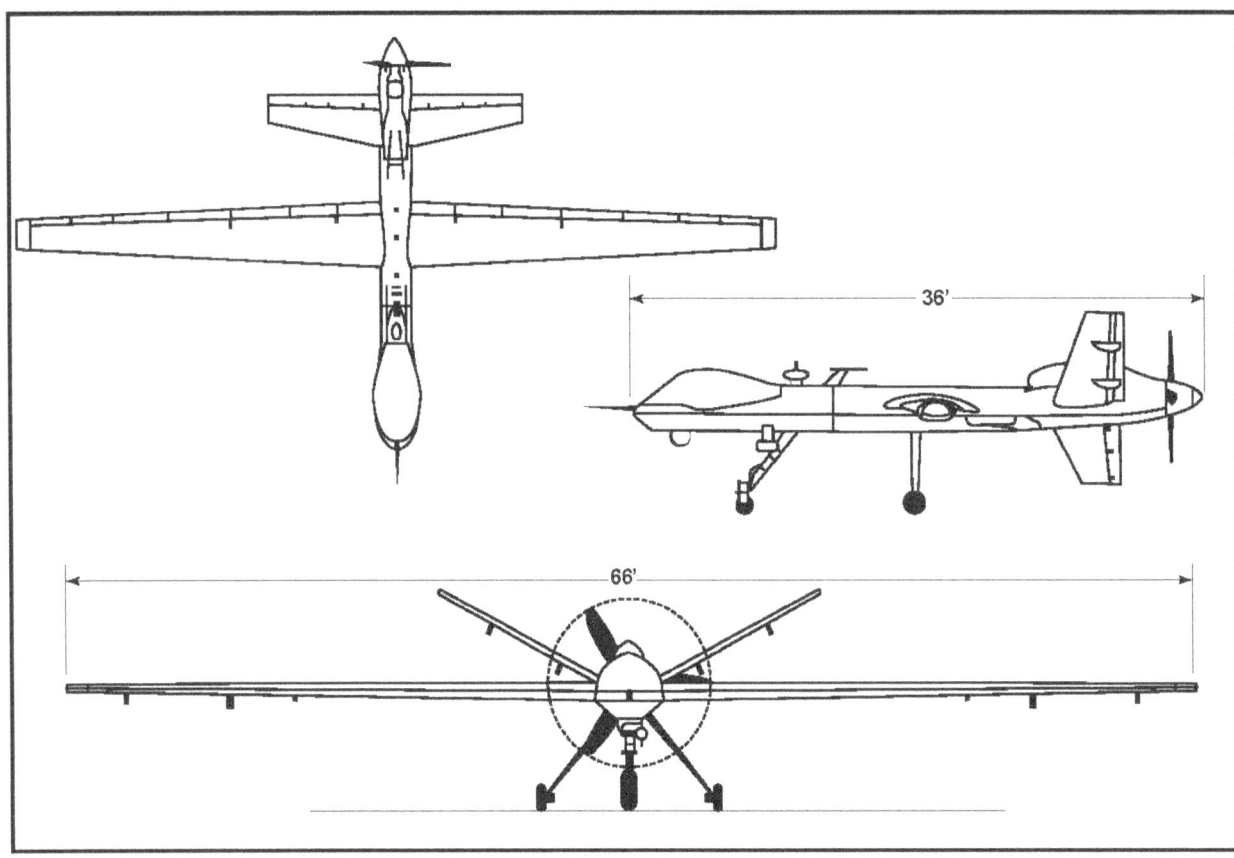

General Atomics

Item	Weight/Dimension
Gross Takeoff Weight	10,000 lbs
Empty Weight	3,700 lbs
Maximum Fuel Load	4,000 lbs
Nose Payload	800 lbs
Max Payload with Max Fuel	2,300 lbs
Max Payload	3,800 lbs
Max Landing Weight	6,500 lbs
Inboard Wing	1,500 lbs each
Middle Wing	650 lbs each
Outboard Wing Stations	150 lbs each
Wing Span	66 feet
Wing Area	256 square feet
Wing Aspect Ration	16:1
Wing Root Chord	65.2 inches
Wing Tip Chord	28.8 inches
Overall Fuselage Length	36 feet
Fuselage Height	12.5 feet

provide 2,000 strain measurements in real time, allowing researchers to study changes in the shape of the wings during flight. The fibers are small enough that they have no affect on aerodynamic lift and drag. The sensor system weighs just a few pounds and the fibers are thin enough that future versions could be embedded within a composite wing structure. Such sensors have numerous applications for future genera-

tions of aircraft and spacecraft. They could be used, for example, to enable adaptive wing-shape control to make an aircraft more aerodynamically efficient for particular flight regimes.[29]

Flying the Ikhana UAS

Using the newly built Ikhana aircraft, General Atomics personnel trained the NASA Ikhana crewmembers – pilots, systems monitors, and maintenance technicians – at the company's Gray Butte facility. This effort culminated in a milestone flight on June 23, 2007. Herman Posada, flying Ikhana from a Ground Control Station at Gray Butte, initiated takeoff and flew the aircraft into restricted airspace bordering Edwards Air Force Base. He then handed off control to Mark Pestana in the GCS at NASA Dryden, who landed the aircraft on the main runway at Edwards and taxied it to the NASA parking ramp.[30]

[29] Jay Levine, "Measuring up to the Gold Standard," *X-tra*, NASA Dryden Flight Research Center, Edwards, CA, 2008.

[30] Jay Levine "No one on board – Ikhana pilots fly aircraft from the ground," *X-tra*, NASA Dryden Flight Research Center, Edwards, CA, 2008.

The Ikhana is flown from a ground control station, visible at right. A Ku-band satellite communications system provides data uplink and downlink capability.
NASA

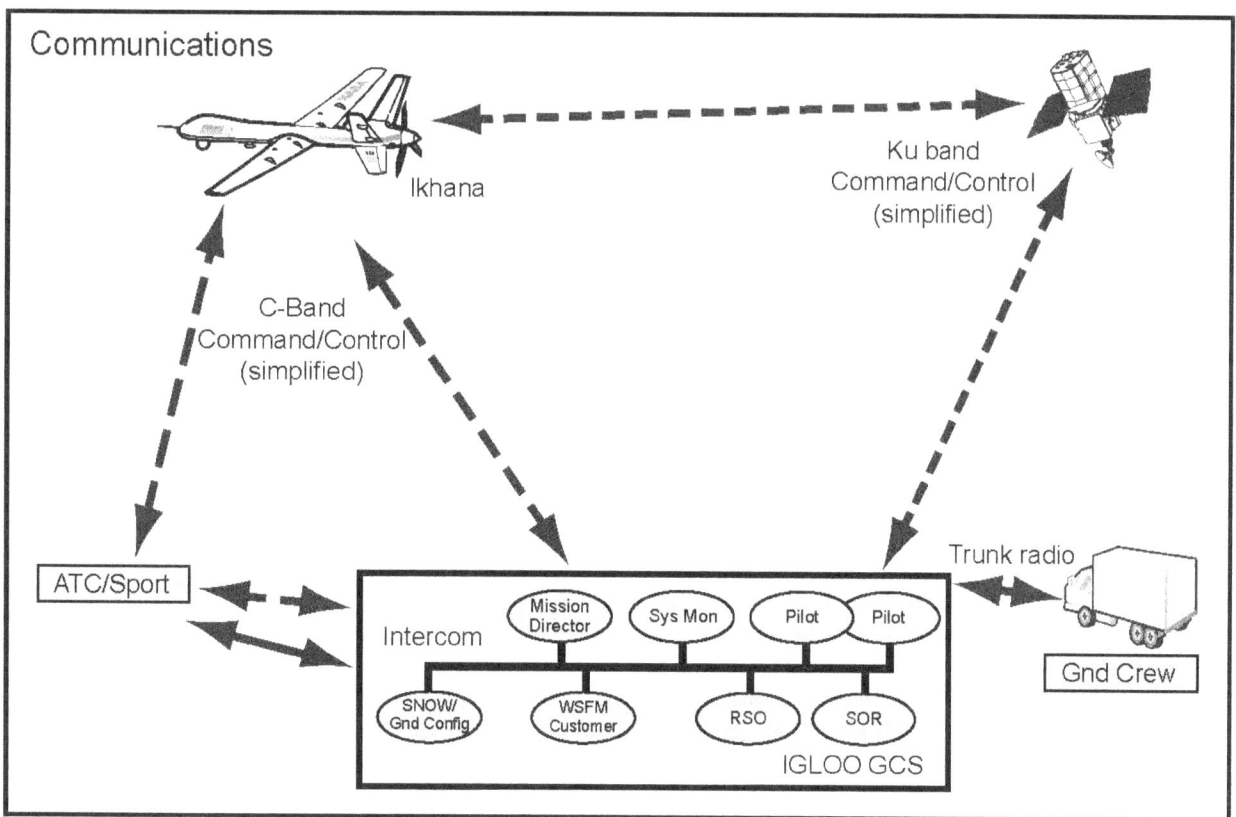

Pilots and mission support personnel control Ikhana from the GCS using direct (C-band) and satellite-relayed (Ku-band) communications. Air-traffic control issues are coordinated with FAA ATC personnel and military controllers (Sport) at Edwards Air Force Base.
NASA

NASA research pilots Herman Posada, left, and Mark Pestana found flying the Ikhana an exciting challenge due to lack of ordinary physical cues and visibility restrictions.
NASA

In order to fly the Ikhana, NASA purchased a GCS and satellite communication system for uplinking flight commands and downlinking aircraft and mission data. The GCS is installed in a mobile trailer and, in addition to the pilot's remote cockpit, includes computer workstations for scientists and engineers. All the aircraft systems are mobile, making the Ikhana ideal for missions conducted from remote sites around the globe.

General Atomics designed the GCS for high mobility and portability. It can be carried aboard a C-130 or any larger transport aircraft. The ground pilot is linked to the aircraft through a C-band line-of-sight (LOS) data link at ranges up to 150 nautical miles. A Ku-band satellite link allows for over-the-horizon control. A remote video terminal provides real-time imagery from the aircraft, giving the pilot limited visual input.[31]

Two NASA pilots, Herman Posada and Mark Pestana, were initially trained to fly the Ikhana. Posada had 10 years of experience flying Predator vehicles for General Atomics before joining NASA as an Ikhana pilot. He soon discovered he had to adapt to differences between the earlier models with single-string avionics and the advanced version with triple-redundant avionics and digital electronic engine controls.

"I think it's just experience," said Posada. "It's not a perfect system, but there are ways to make it work right."[32]

The GCS cockpit features pilot and payload operator stations, the latter of which also serves as a co-pilot station with redundant controls. CRT screens provide a heads-up display with options for a military-style data presentation or one more like what might be found in a civilian light aircraft (referred to by pilots as the "Cessna display"). Only one pilot at a time can fly the aircraft, although the co-pilot assists during the demanding takeoff and landing phases and can issue commands to configure the various aircraft systems as necessary. Unlike in a conventional airplane, control inputs are made with a keyboard and joystick. When one pilot needs to take a break, another can take over so that there is always a fully alert crewmember at the controls.

The entire experience of flying the Ikhana remotely is very different compared to conventional flying because the UAS pilot lacks such physical cues as visibility, motion, sound, feel, and even smell. Pestana, who has over 4,000 flight hours in numerous aircraft types, had never flown a UAS prior to his assignment to the Ikhana project. He found the experience an exciting challenge to his abilities because the lack of vestibular cues and peripheral vision hindered his situational awareness and eliminated his ability to experience such sensations as motion and sink rate.

"It was like I had lost four of my five senses," he said of his experience in the ground cockpit. "Your vision is limited because there is only a single camera for forward visibility, you can't hear the engine, feel the aircraft's motion or acceleration, or smell a fuel leak or an electrical fire."[33]

The flight controls are also markedly different from those of a conventional aircraft. The pilot's station includes a set of rudder/brake pedals, engine

[31] "Ground Control Stations Fact Sheet," General Atomics Aeronautical Systems Company, San Diego, CA, 2007.

[32] Interview with Herman Posada, NASA Dryden Flight Research Center, Sept. 19, 2008.

33 Interview with Mark Pestana, NASA Dryden Flight Research Center, Aug. 13, 2008.

Herman Posada, right, and Mark Pestana seated in the GCS pilot's stations. One pilot flies the aircraft while the other assists with communications and checklists, and serves as a relief pilot when necessary.
NASA

throttle, propeller controls, and a control stick grip but the similarities end there. On a traditional aircraft, control stick deflection to change pitch or bank results in a steady rate of change. On the Ikhana, stick deflection sets a specific fixed pitch or bank angle. Instead of a standard control panel, the pilot uses two systems display screens to access more than 60 pages of data.[34]

"Instead of physical switches – toggle switches or dials – you're using a keyboard and trackball and pulling down menus like you would on your personal computer to activate systems," Pestana explained. "Understanding where all of these system controls are located, and finding the right screen display to access the controls, is challenging."[35]

Posada described an Ikhana flight as "hours of

On June 23, 2007, Herman Posada flew the Ikhana from Gray Butte into restricted airspace bordering Edwards Air Force Base. He then passed control to Mark Pestana in the GCS at NASA Dryden Flight Research Center.
NASA

[34] Jay Levine, "No one on board – Ikhana pilots fly aircraft from the ground."

[35] Ibid.

Above, the Ikhana is guided to its first landing at Edwards by pilot Herman Posada. The flight marked a milestone as Ikhana operations were moved from Gray Butte to Dryden.
NASA

At left, the GCS includes stations for systems monitors and project scientists. Computer terminals display data downlinked from the vehicle.
NASA

boredom punctuated by a couple of moments of sheer terror during takeoff and landing." The demands of these maneuvers place a considerable workload on the pilot.[36]

"There's a lot of stuff you're looking at while working the radios and checklists. It's a little too much for one (pilot). You need an extra set of eyes because sometimes you're drowning in information. Having other people say your speed is high or fast, or telling you to watch your sink rate is important," he explained, emphasizing that flying the Ikhana is a team effort. "There are a lot of people on the team. Without their vital support, we couldn't get the airplane in the air."[37]

Endurance is a key advantage in using a UAS for a variety of science missions. The Ikhana's range capabilities and the fact that flight crews can be changed without the aircraft returning to base allow the vehicle to remain aloft for more than 20 hours during a single mission.

[36] Interview with Herman Posada, Sept. 19, 2008.

[37] Jay Levine, "No one on board."

"For example, " said Pestana, "atmospheric scientists prefer to have continuous data collection over a full day's cycle, where the presence or absence of sunlight may drive chemical reactions in the atmosphere that affect weather and climate."[38]

These characteristics were at the heart of efforts to establish a project to apply UAS technology to fighting wildfires.

Forward-looking cameras provide visual information for situational awareness. A head-up display overlays information regarding airspeed, altitude, attitude, engine performance, and landing-gear status.
NASA

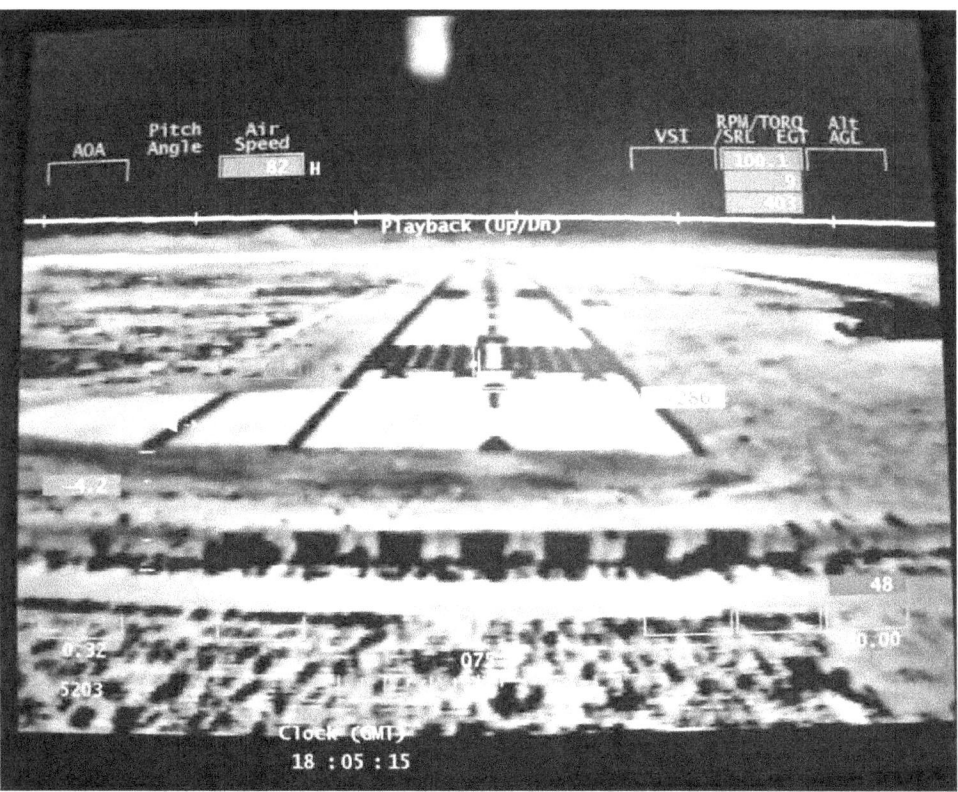

An infrared camera provides visibility during nighttime operations, allowing the pilot to land the vehicle safely in the dark.
NASA

[38] Interview with Mark Pestana, Aug. 13, 2008.

19

CHAPTER TWO
CHARIOTS OF FIRE

One obvious application of UAS-based remote sensing technology is in the field of emergency response to wildfires. In order to combat fire in rugged wilderness terrain, first responders need as much information as possible, as quickly as possible. That drives a need to use airborne sensors with a wide field

Previously, on older systems, sensor data had to be physically downloaded and processed after landing. This meant the information was hours old by the time it reached emergency response personnel. Ikhana's high-bandwidth data link enables fire information to be transmitted to incident commanders within minutes, a

Smoke and ash can obscure firefighters' views in densely forested areas. Airborne thermal-infrared sensors provide a bird's-eye view while penetrating the haze layer.
National Park Service

of view (several miles) that can penetrate smoke and haze obscuring the firefighter's view of thick forests or brush-filled canyons and rugged terrain. While the sensors themselves could be mounted in any number of different aircraft, flying such a mission poses a risk to aircraft and crews. Hazards include obscured visibility due to smoke, night operations, and turbulence from rising columns of hot air. Use of unmanned aircraft mitigates these risks while delivering timely information.

vital capability when high winds are carrying embers over long distances in rugged terrain.

In September 2001, scientists from Ames Research Center led a demonstration of the Altus II UAS as a remote sensing platform to gather thermal data over wildfires and relay real-time information via satellite to fire management personnel on the ground. The First Response Experiment, or FiRE, demonstrated the combined use of a multi-spectral scanner, satellite image data telemetry, near real-time image geo-rectification,

Crews set a backfire to stop the Poomacha fire from advancing westward. Airborne-sensor data, delivered in near real-time, allow incident commanders to deploy firefighting resources more effectively.
FEMA/Andrea Booher

and rapid data dissemination over the Internet to disaster managers.[39]

Dr. Steven S. Wegener, senior research scientist at NASA-Ames (now with the Bay Area Environmental Research Institute) spearheaded the FiRE demonstration missions on the Altus II UAS. Vincent G. Ambrosia, senior research scientist at California State University, Monterey Bay, was a fire science team led on the FiRE effort with a primary interest in demonstrating UAS imaging capabilities for supporting wildfire assessments. The Ames wildfire science team's interest came in response to an upward trend in the number and severity of wildfires, a key component of changing ecosystems and climate impacts. "We began looking at the science of fires," Ambrosia said, "and then started working on improving the capabilities of airborne sensors to better observe and study fires."[40] The team

at Ames developed the Airborne Infrared Disaster Assessment System (AIRDAS) sensor package to fill that critical fire sensor niche in the early to mid-1990s. The AIRDAS was flown on numerous missions on manned platforms until Wegener developed the idea of integrating the AIRDAS on the Altus II to demonstrate the efficacy of using unmanned systems and sensors for science and applications support missions.

The AIRDAS is a four-channel line-scan instrument designed to measure the thermal signature of wildfires or other natural and manmade disasters, accurately resolving fire intensities as high as 600 degrees C. The instrument collects data in four filterable electromagnetic channels to provide various types of information for analysis. Band 1 (visible) is suitable for monitoring smoke plumes as well as distinguishing cultural and topographic features. Band 2 provides information for analysis of vegetative composition and very hot fire fronts while penetrating most associated

[39] Vincent G. Ambrosia, "Remotely Piloted Vehicles as Fire Imaging Platforms: The Future is Here!" See also *Wildfire Magazine*, article available online at http://geo.arc.nasa.gov/sge/UAVFiRE/completeddemos.html, May-June 2002, accessed June 10, 2009.

[40] Interview with Vincent G. Ambrosia, NASA Dryden Flight Research Center, Edwards, CA, Aug. 13, 2008.

Vincent Ambrosia headed a team that developed an airborne sensor package to collect imagery of wildfires from an unmanned aircraft, process the data on board, and transmit them to users in near-real time.
NASA

smoke plumes. Band 3 (mid-infrared spectrum) is designed for estimating high-temperature conditions. Band 4 is used to collect thermal data on Earth ambient temperatures, lower temperature soil heating conditions behind fire fronts, and minute temperature differences in pre-heating conditions. The system operates at five to 24 scans per second with a digitized swath width of 720 pixels in the cross-track direction as data is acquired continuously in the along-track direction. The scanning optics have a 108-degree field of view in the cross-track direction and an instantaneous field of view of 2.62 milliradians. These parameters provide a ground resolution of 8.0 meters from an altitude of 10,000 feet above the ground. Fires smaller than 8.0 meters are detectable through calibration.[41]

Image data is downlinked from the aircraft and geo-rectified so it can be almost immediately integrated into a map base. Typically, the information is available to firefighters within five to 10 minutes of

acquisition. It can sometimes take as much as 30 minutes because the system can either acquire or downlink data, but not simultaneously. "Real-time data is critical in a disaster such as a fire incident," said Ambrosia. "Responders need to know where the fire is right now and where it is going."[42]

FiRE demonstration

The FiRE demonstration took place Sept. 6, 2001, at El Mirage. Disaster managers and fire management personnel were on hand to view the demonstration and participate in evaluation of products and procedures. Technicians initiated aircraft system and payload checks one hour before scheduled takeoff time, with

[41] Vincent G. Ambrosia, "Remotely Piloted Vehicles as Fire Imaging Platforms: The Future is Here!"

[42] Interview with Vincent G. Ambrosia.

The Altus II was used for the First Response Experiment (FiRE) at El Mirage in September 2001. A sensor package on board successfully transmitted data to scientists at NASA Ames Research Center at Moffett Field.
NASA

"hands off" all systems about 30 minutes later. Altus II took off immediately after a controlled burn was ignited adjacent to the runway. After reaching an altitude of 6,000 feet above the ground, controllers flew the vehicle in a racetrack pattern over the burn area for a total of five data-collection passes. Technicians on the ground geo-rectified the imagery and passed the data to fire managers via computer network within 10 minutes of receiving them. Exceeding expectations, the Altus II UAS had launched, attained mission altitude, completed five data-collection passes, telemetered AIRDAS data over the horizon via satellite to Ames, and data were geo-rectified and distributed to fire managers via Internet within an hour of takeoff.[43]

Success of the FiRE demonstration was the result of close cooperation among federal and state agencies and private industry. General Atomics personnel designed, built, and flew the Altus II vehicle and performed systems integration. NASA scientists developed and supplied the AIRDAS thermal imaging scanner. Remote Satellite Systems Inc. of Santa Rosa, Calif., provided a NERA World Communicator M4

portable satellite telephone terminal and antenna for telemetry. GA-ASI personnel integrated the telemetry system into the airplane's fuselage after NASA technicians modified the equipment for remote aircraft operations. Terra-Mar Resource Information Services of Mountain Ranch, Calif., performed near real-time image geo-rectification.[44]

In the wake of the Altus II success, the FiRE project team focused on further advancements in UAS technology, payload capabilities, telemetry, and information processing for disaster management. The next phase of research, development, and demonstration involved the Altair UAS with plans for long-duration missions covering actual wildfires throughout the Western United States.

In 2003, NASA funded the Wildfire Research and Applications Partnership (WRAP), a five-year project

[43] Ambrosia, "Remotely Piloted Vehicles as Fire Imaging Platforms: The Future is Here!"

[44] Ibid.

to foster collaborative efforts among NASA and U.S. Forest Service personnel to develop and demonstrate technologies for collecting and sharing data on wildfires. Evolving technologies were exploited to improve information content and timeliness of data dissemination to fire managers. Plans to accomplish these objectives evolved into the Western States Fire Mission (WSFM).[45]

Sensor development

The Autonomous Modular Sensor (AMS), developed by a team at NASA Ames was key to the success of the WSFM. This was not, however, the first use of NASA resources for imaging wildfires.

In 1971, the Air Force loaned two U-2C airplanes to NASA Ames for use as Earth Resources Survey

NASA operated U-2C and ER-2 aircraft in a variety of environmental research projects, including collection of wildfire imagery. The greatest disadvantages included pilot endurance and the length of time required to retrieve and process data.
NASA

[45] Philip Hall, Brent Cobleigh, Greg Buoni, and Kathleen Howell, "Operational Experience with Long Duration Wildfire Mapping UAS Missions over the Western United States," presented at the Association for Unmanned Vehicle Systems International Unmanned Systems North America Conference, San Diego, CA, June 2008.

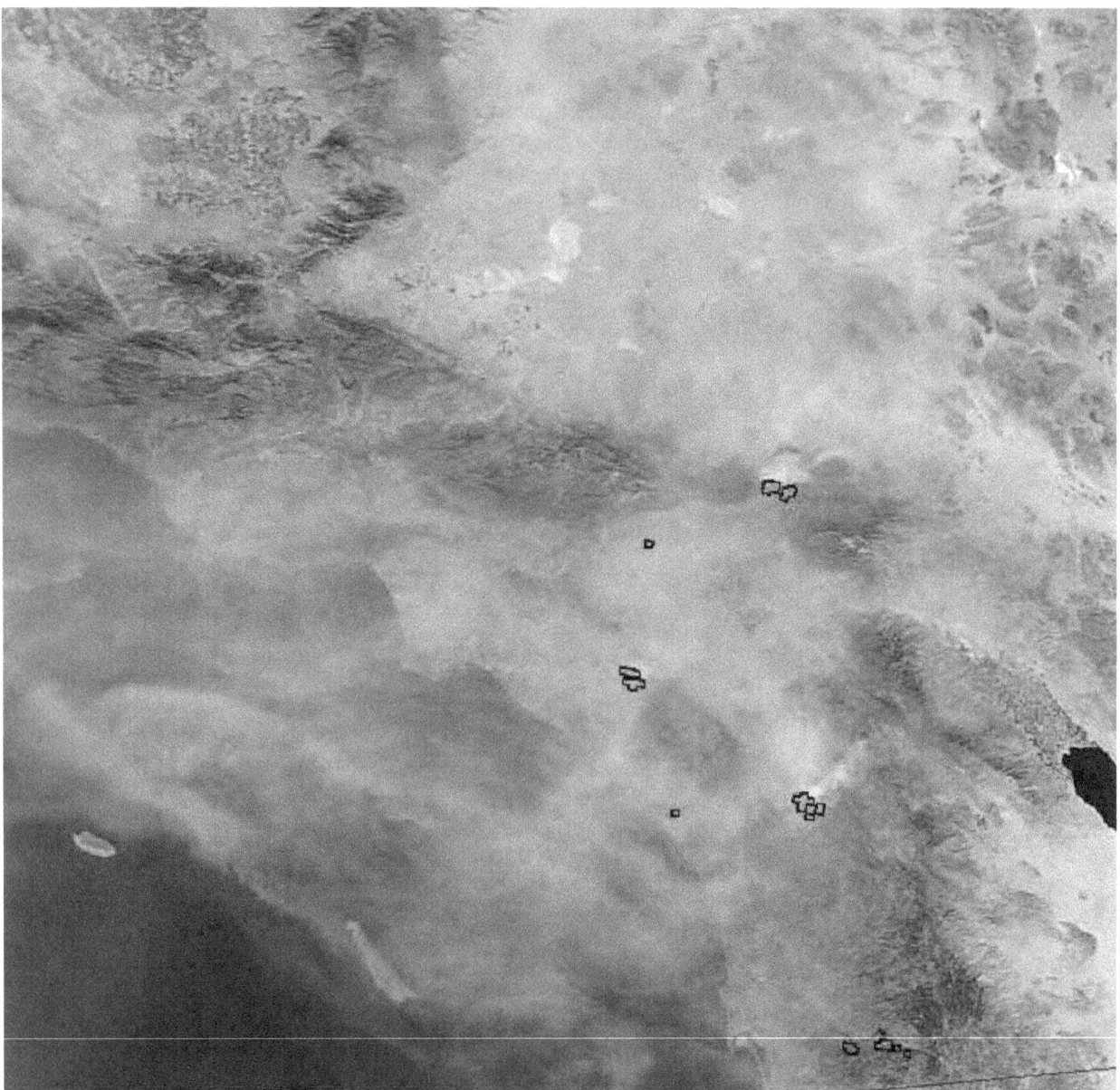

The EOS-Terra satellite captured this MODIS image of California wildfires on October 25, 2007. Ikhana imagery, acquired at lower altitudes, has greater resolution.
NASA

Aircraft. Visual and infrared photography from these platforms, taken at altitudes in the vicinity of 65,000 feet, showed previously unknown access routes to advancing fire fronts, indicated ideal placement of firebreaks, and identified dangerous terrain where firefighters might become trapped. Experience with U-2 imagery has shown the data helped emergency response personnel contain fires sooner with less manpower and equipment than would have been possible otherwise. The biggest disadvantage, however, was the length of time necessary to retrieve and process the film and deliver the images to incident commanders.[46]

In 1981 and 1989 NASA acquired two ER-2 aircraft, a larger and more capable version of the U-2 that offered a marked improvement in payload capacity, endurance, and range. The ER-2 was capable of carrying NASA's Moderate Imaging Spectroradiometer – MODIS – Airborne Simulator (MAS) instrument, a modified Daedalus Wildfire scanning spectrometer that provides spectral information similar to that provided by MODIS launched on the EOS-AM satellite.

[46] Jay Miller, *Lockheed U-2*, Aerofax Inc., Austin, TX, 1983.

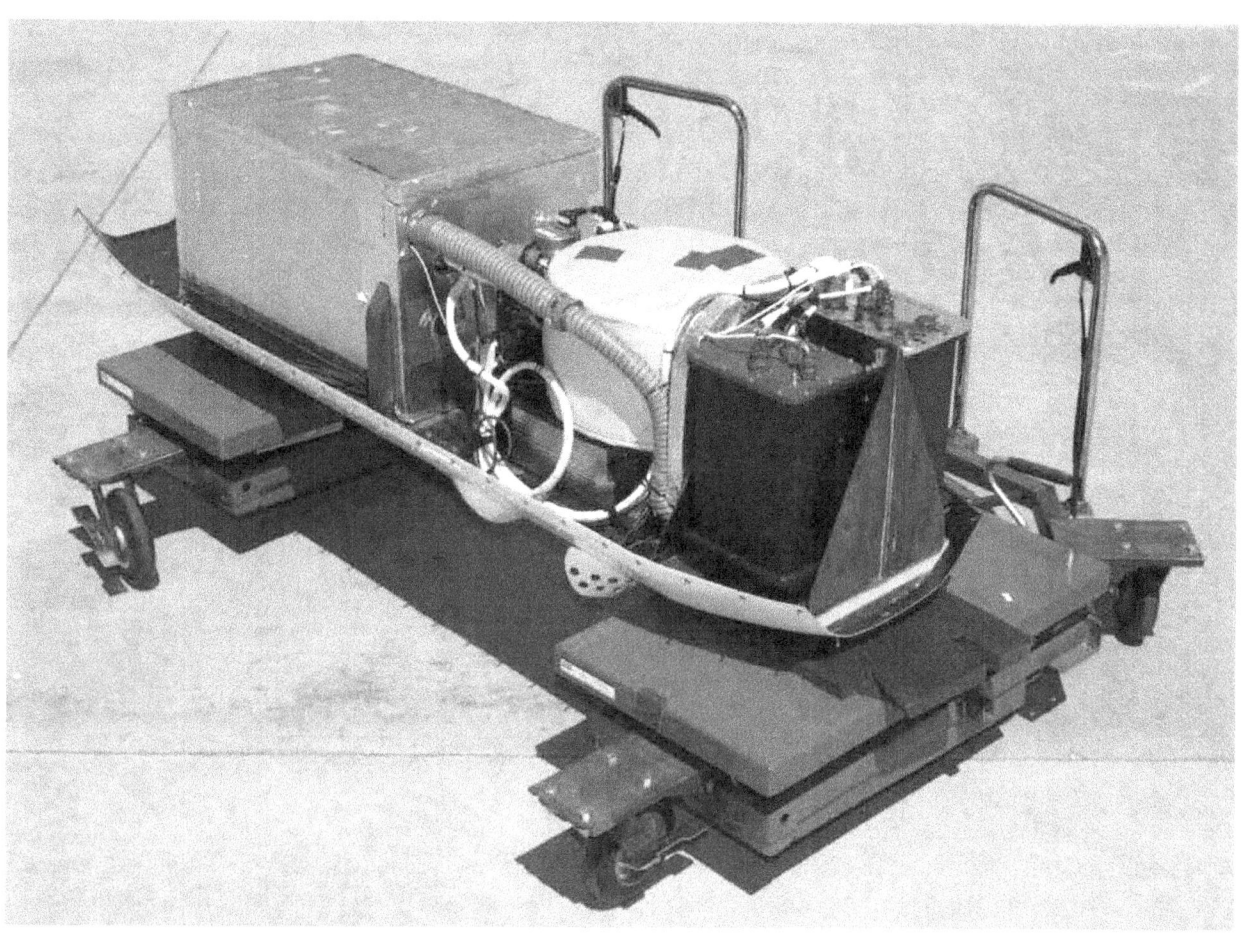

Instruments that make up the Autonomous Modular Scanner include data system computers, scan-head, precision navigation subsystem, power distributor, and control systems.
NASA

The Wildfire Spectrometer was delivered to Ames in April 1991. A single visible-band channel was added and several infrared spectral channels were altered to configure the instrument for the FIRE Cirrus-II experiment. In January 1992 the modified Wildfire sensor was then further modified to become MAS. Although the instrument is a 50-band spectrometer, the digitizer was configured for each mission to record a pre-selected group of 12 spectral bands during the flight. For most of these missions the digitizer was configured to record four 10-bit channels and seven 8-bit channels. A 50-channel digitizer capable of recording all 50 spectral bands at 12-bit resolution became operational in January 1995.[47]

NASA officials recognized that a sensor package that could be mounted on a remotely piloted aircraft would provide greater flexibility, and tasked researchers at Ames with developing instrumentation and demonstrating that scientifically valuable data could be gathered using UAS technology.

Subsequently, a team of scientists began working on an upgraded version of the Daedalus AADS-1268 Thematic Mapper Simulator (TMS) – a digital multispectral scanner flown aboard the ER-2 – but with a lighter, smaller, more autonomous sensor package and lower electrical power requirements. The team's engineering goals included development of an accurate scientific sensor that would be compatible with a remotely piloted aircraft and capable of providing data to users as quickly as possible. The sensor was designed to be modular so that the 12-band Wildfire spectrometer could be replaced with one of two others, which featured different spectral bands optimized for

[47] "First ISCCP Regional Experiment (FIRE) Cirrus 2 NASA ER-2 Moderate Resolution Imaging Spectroradiometer (MODIS) Airborne Simulator (MAS) (FIRE_CI2_ER2_MAS) Langley DAAC Data Set Document," Atmospheric Science Data Center, NASA Langley Research Center, Hampton, VA, http://eosweb.larc.nasa.gov/GUIDE/dataset_documents/base_fire_ci2_er2_mas_dataset.html, 1996, accessed June 10, 2009.

The Altair carried the sensor pod on a ventral centerline pylon. Sensor data is autonomously processed with georectified topographical information to create a fire-intensity map.
NASA

ocean and atmospheric detection, depending on mission requirements. The sensor pod was originally configured for ventral carriage on the Altair's fuselage centerline. For Ikhana missions, the pod was mounted beneath the left wing.

The MODIS was used as a starting point for the AMS Wildfire sensor since data from the scanner's spectral bands could be input into a fire detection algorithm used by the Forest Service and other agencies. The new sensor system needed the capability to provide general Earth resources imaging, and input for the fire detection algorithm, and would need to be capable of delivering accurate, readily interpreted images of affected areas in near real-time, so that firefighting resources can be efficiently deployed.

The result was the AMS, a line scanner with a rotating elliptical mirror that provides a cross-track scan. Incoming light reflected off the input mirror is sent through a telescope, which collimates the beam. The light then enters a 12-band spectrometer assembly. Eight wavelength bands cover the spectral range from visible to the near infrared. These bands are dispersed through a prism and collected by an array of silicon photodiodes. Two additional near-IR bands are routed through filters and dichroic mirrors to two thermoelectrically cooled indium gallium arsenide detectors. Dichroic mirrors and a dual-band bandpass filter select the final two mid-infrared

and far-infrared bands, which are specially defined for wildfire applications. The bandpass filter sits in front of a sandwiched detector with an indium-antimony detector atop a mercury-cadmium telluride detector, cooled by a Stirling cycle cryogenic cooler. Data from these bands are processed in the fire detection algorithm. Onboard blackbody sources calibrate each detector well enough that the algorithm can report the temperature of the detected fires up to a temperature of 1,000° C (1,832° F) with an accuracy of 0.5° C (16.9° F). The sensor can also provide a vector map of hot spots.

While operating, output from the detectors is digitized into 716 16-bit cross-track pixels. For wildfire imagery the pixel angular resolution is about 100 ft/pixel for data acquired from 40,000 feet, providing an appropriate tradeoff between resolution and coverage. Digitized data is combined with navigational and inertial sensor data to precisely determine the location and orientation of the sensor. In addition, the data is autonomously processed with georectified topographical information to create a fire intensity map.

The AMS weighs about 264 pounds, including heating and cooling equipment, optical assembly, digitizer, and processor. With blackbody calibration sources and detector heaters on, the system can use up to 50 amps at 28 volts DC. Data collected by this small package are processed on board the aircraft to provide a finished

General Atomics Aeronautical Systems Inc. supplied pilots and sensor operators for the Altair Western States Fire Missions. NASA

product formatted according to a geographical information systems standard, which makes it accessible with commonly available programs such as Google Earth.

The AMS records all 12 channels on board so that all data is available for post-flight processing. Generally, during a mission, only three channels of desired information are downlinked in near real-time. The channels selected depend on the needs of the user – day/night imagery, active fire data, post-burn data, etc. Data telemetry is downlinked via a Ku-band satellite communications system. After quality-control assessment by scientific personnel in the GCS, the information is transferred to NASA Ames and then made available to remote users via the Internet.[48]

The AMS payload also drapes fire information over three-dimensional geographic data collected by a specially modified radar system that flew on board the space shuttle Endeavour during an 11-day mission in February 2000. The Shuttle Radar Topography Mission (SRTM) obtained elevation data on a near-global scale to generate the most complete high-resolution digital topographic database of Earth. The NASA Jet Propulsion Laboratory released SRTM data for public use through the Internet.[49]

Altair Western States Fire Missions

Goals for the 2006 WSFM included demonstrating the capabilities of a long-duration UAS to collect infrared imagery of wildfires in the Western United States, west of the Rocky Mountains and between the Mexican and Canadian borders, and to disseminate the data in near real-time to firefighters in affected areas. Partners from NASA Ames and the Forest Service addressed scientific and mission goals while those at NASA Dryden were responsible for operational aspects of flying a remotely piloted aircraft in the national airspace. General Atomics was responsible for constructing a pod to hold the AMS and integrating the pod and sensor package onto the aircraft.[50]

Three operational objectives were critical to the mission. First, the UAS had to remain aloft for long durations to image multiple fires throughout the region of interest. Second, NASA had to develop a sensor and associated equipment and systems capable of collecting, processing, and delivering information to firefighters and incident commanders within 10 minutes of collection. Finally, the team had to demonstrate the ability to safely operate a UAS in the national airspace using the same FAA procedures required of other aircraft. NASA officials conducted mission planning and were responsible for obtaining FAA approval for flights in the national airspace. General Atomics provided pilots and sensor operators.[51]

In early 2006, NASA leased time in the Altair UAS from General Atomics for the Western States

[48] Richard Gaughan, "An autonomous sensor developed by NASA proves its worth in firefighting," R&D Daily, Issue 0701, January 2007.

[49] Philip Hall, Brent Cobleigh, Greg Buoni, and Kathleen Howell, "Operational Experience with Long Duration Wildfire Mapping UAS Missions over the Western United States."

[50] Gregory P. Buoni and Kathleen M. Howell, "Large Unmanned Aircraft System Operations in the National Airspace System – the NASA 2007 Western States Fire Missions," AIAA-2008-8967, American Institute of Aeronautics and Astronautics, The 26th Congress of International Council of the Aeronautical Sciences, Anchorage, AK, 2008.

[51] Ibid.

Equipped with a pod-mounted infrared-imaging sensor, the Altair UAS aided fire-mapping efforts over wildfires in central and southern California in late 2006.
NASA

Fire Mission. The NASA Dryden team was responsible for mission planning and obtaining a Certificate of Authorization (COA) from the FAA, granting permission for Altair to overfly fire incident areas within the national airspace.[52]

Working with the newly created FAA Headquarters Unmanned Aircraft Systems Program Office, the NASA team made significant progress toward completing a COA application for flights to capture data over wildfires in any of the 11 western states: California, Oregon, Washington, Nevada, New Mexico, Arizona, Idaho, Montana, Utah, Wyoming, and Colorado. By September, however, it became clear that COA approval would not be possible before the close of the 2006 wildfire season. In order to get the AMS into the air as soon as possible and begin data collection, the team requested a COA for a more limited operating area. On Oct. 19, a COA valid through Dec. 1 was granted that allowed Altair to be flown in areas directly adjacent to restricted airspace normally used Dryden-based NASA research missions.[53]

Using this authorization on Oct. 24, 2006, the team flew the Altair on a 23-hour mission over a controlled burn near Yosemite National Park, Calif., at an altitude of 43,000 feet. The data confirmed satellite-based fire detection information and the AMS payload also observed a second nearby wildfire. Following this significant milestone, the Western States Fire Mission team stood down, removed the sensor pod from Altair, and began preparing for the 2007 wildfire season.

Plans changed quickly when an arsonist set a small fire near Cabazon, Calif., that grew to devastate more than 40,000 acres, destroy 34 homes, and cause the deaths of five firefighters. California governor Arnold Schwarzenegger requested emergency support for remote sensing over the burn area of the Esperanza fire, as it came to be known. The Altair team responded to the crisis by reinstalling the pod, rewiring the Altair aircraft, planning the mission, and requesting a COA Amendment.[54]

[52] Ibid.

[53] Ibid.

[54] Jamie Wilhite, Robert Navarro, and Brent Cobleigh, "Altair Western States Fire Mission," 2006 Engineering Annual Report, pp. 45-47, NASA Dryden Flight Research Center, Edwards, CA, August 2007.

The Altair was flown over a controlled burn near Yosemite National Park as well as a nearby wildfire to confirm satellite-based fire-detection information and to test the AMS sensor.
NASA

NASA planners submitted the request on Oct. 27, asking for an extension of the Yosemite COA that would allow a flight over the Esperanza fire. Fortunately, the FAA had created a process that allowed the extension to be granted that evening. Altair was ready to fly again.[55]

During a 16-hour mission, the team delivered some 100 visible and infrared images in near real time to incident commanders along with more than 20 data files with detailed maps of the fire's perimeter. The Incident Command Team studied the thermal data and used the technology to develop an Incident Action Plan that was distributed to firefighters the next morning. The information helped responders better understand the location and movement of the fire and distribute resources accordingly.[56]

[55] Gregory P. Buoni and Kathleen M. Howell, "Large Unmanned Aircraft System Operations in the National Airspace System – the NASA 2007 Western States Fire Missions."

[56] Daniel Berlant, "Unmanned aircraft is latest firefighting tool," Communique, California Department of Forestry and Fire Protection, Sacramento, CA, http://www.fire.ca.gov/communications/downloads/communique/2007_winter/unmanned.pdf. Winter 2007, accessed June 10, 2009.

NASA scientists in the GCS at Grey Butte collected AMS images downlinked from the Altair and retransmitted them to the wildfire Collaborative Decision Environment (CDE) at Ames via the Internet. The images were processed on board the Altair with geo-rectification software to precisely overlay them onto Google Earth global map and satellite imagery. On-site fire commanders could then view these images via Internet connections. During the mission, the Altair also carried several instruments to collect atmospheric data. These included an Argus tunable laser diode spectrometer developed at Ames, a NOAA gas chromatograph ozone photometer, and Dryden's Research Environment for Vehicle-Embedded Analysis on Linux (REVEAL) instrument – a programmable gateway between onboard instruments and wireless communication paths to and from the aircraft. The Argus sensor was used to collect carbon monoxide measurements that helped atmospheric scientists better understand the dynamics of the atmosphere over a vertical range up to 35,000 feet. Besides providing valuable scientific data, it gave researchers information on the health of the instrument itself on long-duration, high-altitude

The ground control station includes consoles for two pilots and positions for scientists and engineers.
NASA

flights. The Argus also validated information previously collected with NASA's Aura satellite.[57]

"As a fire department and especially a wildland one, we're going to look at the technology that's out there," said Riverside County fire captain Julie Hutchinson. "The sooner we get information to the ground forces and fire managers, that makes a difference. That's a huge thing for us."[58]

Ikhana Western States Fire Missions

Building on the experience with the Altair, NASA acquired the Ikhana UAS for a second series of Western States Fire Mission flights. Team members worked hard to complete a significant number of training, engineering, and operational milestones in preparation for operational missions scheduled to begin in August 2007. These included training for Dryden pilots, crewmembers, and technicians, as well as integration and testing of the GCS. The aircraft was modified to carry

the sensor pod on a wing pylon, and technicians integrated and tested all associated hardware and systems. Management personnel at Dryden performed a flight readiness review to assure that all necessary operational and safety concerns had been addressed. Finally, planners had to obtain the necessary COA to allow the Ikhana to operate in the national airspace.[59]

NASA flight operations were conducted as a coordinated effort with the FAA's service areas, Air Route Traffic Control Centers (ARTCC), and Unmanned Aircraft Program Office (UAPO). Significant hurdles included flight approval from the UAPO and a NASA Dryden safety board. Each review required detailed descriptions of mission plans, procedures, and contingencies. Additionally, the Dryden safety review included all engineering and training issues. The Ikhana's initial flights originated from Gray Butte but on June 23, 2007, the aircraft was flown to Dryden – its permanent home.[60]

[57] Jamie Wilhite, Robert Navarro, and Brent Cobleigh, "Altair Western States Fire Mission."

[58] Berlant, "Unmanned aircraft is latest firefighting tool."

[59] Philip Hall, Brent Cobleigh, Greg Buoni, and Kathleen Howell, "Operational Experience with Long Duration Wildfire Mapping UAS Missions over the Western United States."

[60] Ibid.

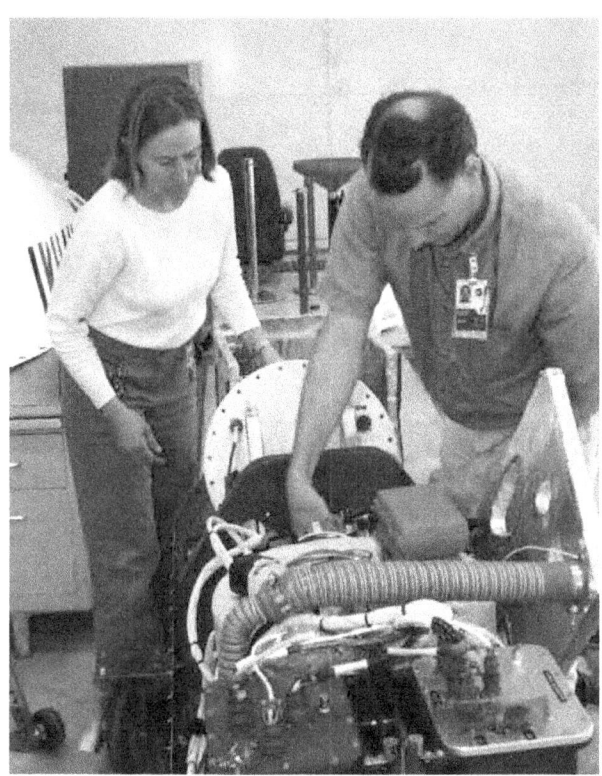

NASA Ames engineers Sally Buechel and Ted Hildum prepare to load the Autonomous Modular Scanner into the Ikhana unmanned aircraft's payload pod.
NASA

Flight planning

Due to the large geographical area involved for each fire mission sortie and the flight schedule flexibility being requested, FAA officials considered the WSFM one of the most complex UAS missions to date. Flexibility was a key issue because the unpredictable nature of wildfires makes it difficult to accurately plan flight tracks more than a few days in advance.

NASA officials submitted the WSFM COA application in February 2007, building on experience gained in the Altair mission planning process. All details of the 2007 mission plans and pertinent information regarding the Ikhana were submitted electronically to the FAA Headquarters UAS Program Office. This was followed by a series of informal meetings between NASA and FAA officials to clarify details of the COA application, mission goals, and operational plans. These meetings gave the FAA ample time to review and discuss the details with the Ikhana team.[61]

"The FAA has been very cooperative in helping to

define ways to achieve our mission objectives while protecting the safety of the national airspace system," said Greg Buoni, lead operations engineer for Ikhana. "Because unmanned aircraft currently have limited ability to see and avoid other aircraft and, in some cases, have lower reliability than a manned aircraft, unmanned flights within the national airspace require a COA and are subject to significant restrictions in their operation."[62]

Plans for the 2007 WSFM called for a total of four to five flights of approximately 12 to 24 hours duration. Because the FAA required flight plans be submitted three business days prior to the planned flight, the NASA team had to submit plans on Monday for flights scheduled for Thursday. Friday and Saturday served as back-up mission days. Crew rest and aircraft maintenance requirements precluded back-to-back missions.

The magnitude and complexity of the WSFM made it necessary for NASA to coordinate with multiple Air Route Traffic Control Centers. To simplify planning, the area of interest was divided into three zones, each containing no more than three ARTCCs. Each mission would be limited to a single zone in order to reduce the number of people needed to coordinate the flight, particularly within the FAA.[63]

Restrictions imposed by the UAPO and NASA safety policy precluded flying the Ikhana over densely populated areas. In addition, the Dryden Safety Office conducted a detailed risk analysis of each proposed mission. This resulted in development of zone maps showing keep-out areas in red and less-densely-populated areas in yellow. The Ikhana could be flown over yellow areas as long as all aircraft systems functioned normally and the vehicle was under the direct control of a remote pilot in the GCS. If a systems failure resulted in degraded aircraft control, mission rules required avoidance of the yellow areas. In the event that direct control was completely lost, a lost-link mission plan was programmed into the Ikhana's autonomous systems to avoid overflying both the red and yellow areas, reducing risk to people and property on the ground.[64]

In April 2007, following discussions with FAA representatives from several affected ARTCCs, Ikhana

---

[61] Ibid.

[62] Hagenauer, "Ikhana UAV Gives NASA New Science and Technology Capabilities."

[63] Ibid.

[64] Ibid.

Ikhana ground crewmen Gus Carreno and James Smith load the thermal-infrared imaging scanner pallet into the Ikhana's underwing payload pod.
NASA

Lead Operations Engineer Greg Buoni and other NASA mission planners developed "backbone" flight tracks for Ikhana that avoided yellow and red areas. Later, "spoke" segments were added to allow the aircraft to fly to the fire zones of interest.

Buoni originally wanted to fly the Ikhana directly from fire to fire as long as it could avoid the yellow and red exclusion zones but thought that the FAA would prefer backbone-and-spoke routes.

"We understood this was not as efficient as we would like," said Buoni, "but we were willing to sacrifice a little efficiency to get an approach the FAA might approve."[65]

The routes were designed to avoid keep-out zones, maximize coverage of areas likely to experience wildfires, and avoid adverse winds and turbulence as much

as possible. Avoiding adverse weather was important for lost-link contingency plans. If the aircraft had to autonomously return to Edwards without a communications link to the GCS, the Ikhana team hoped the aircraft could avoid areas where extreme weather conditions were likely to occur.

In a May 2007 meeting, FAA ARTCC and UAPO representatives rejected the backbone-and-spoke approach and agreed that flying a more direct route from fire to fire would more closely resemble the behavior of other aircraft flying in the national airspace. The COA application retained the backbone-and-spoke route as an example of what a flight plan might look like if there were fires within those areas, and backbones were incorporated into mission plans as necessary.

Missions were planned several days in advance of each flight, with the idea of covering several fires by using point-to-point navigation. Mission altitude was determined by aircraft performance and airspace

[65] Gregory P. Buoni and Kathleen M. Howell, "Large Unmanned Aircraft System Operations in the National Airspace System – the NASA 2007 Western States Fire Missions."

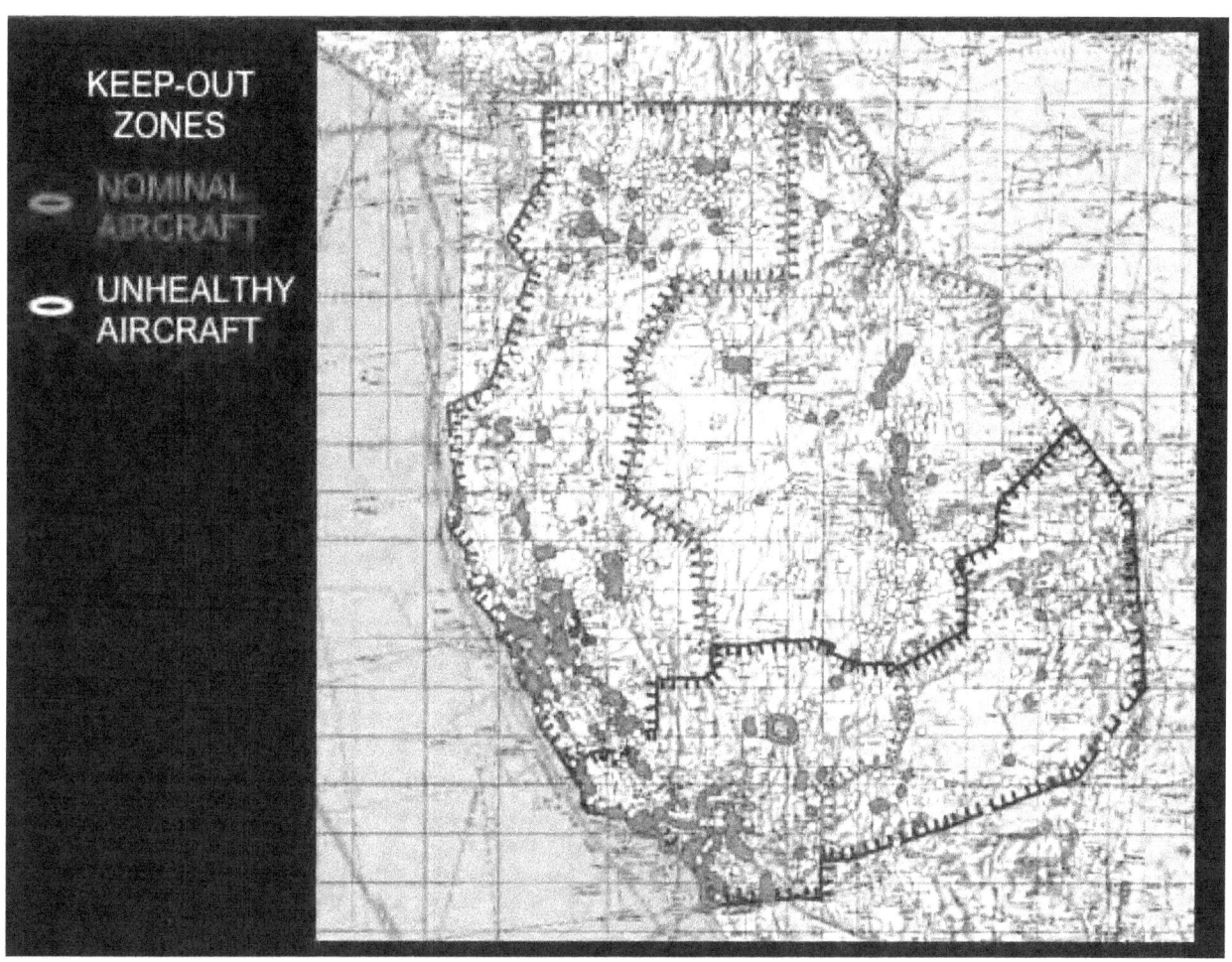

KEEP-OUT
ZONES

NOMINAL
AIRCRAFT

UNHEALTHY
AIRCRAFT

Mission planners established "Keep Out" zones in order to avoid flying the Ikhana over populated areas.
NASA

constraints. Based on the Ikhana's performance characteristics and a desire to avoid bad weather during the summer months, the desired altitude range fell between 35,000 and 45,000 feet. The original 2007 WSFM plan called for flight within Class A (positively controlled) airspace and above Reduced Vertical Separation Minimum (RVSM) airspace. Due to performance limitations and the fact that the aircraft was not certificated to operate in the RVSM band from 29,000 to 41,000 feet, flight was restricted to between 18,000 and 29,000 feet. Flight planners would request an operating altitude of 23,000 feet but real-time altitude changes would be performed as requested by FAA air traffic controllers at any time during the mission.[66]

"See and avoid" capability is a requirement for flight in the national airspace but limitations of the camera system provided on board the Ikhana reduce

the pilot's visual awareness of the aircraft's surroundings. The pilot must rely on air traffic controllers to maintain adequate separation between the UAS and other aircraft. To accommodate these restrictions, the Ikhana was flown through initial climb and final descent within the restricted airspace surrounding Edwards Air Force Base and then transitioned to Class A airspace.

To allow time for dissemination of flight plans to ARTCCs and to brief air traffic control personnel who would be on duty during a WSFM flight, the FAA required 72 hours notice before each mission. Scientific and operational staff had to assemble a flight plan containing navigational waypoints that would provide access to each defined fire area and track lines designed to avoid yellow and red populated areas. Air traffic controllers cleared other aircraft out of the airspace in the vicinity of the fire so the Ikhana could fly track lines and repeat as necessary without the UAS pilot overburdening controllers

[66] Ibid.

"Backbone" routes provided the most direct flight path over each zone of interest while avoiding keep-out areas.
NASA

with radio calls.

The workload imposed on WSFM personnel was intensive. Flight planning began three or four days before a scheduled mission so the paperwork could be submitted to the FAA on a Monday, allowing for a Thursday takeoff. Aircraft preflight preparations usually began several hours before scheduled takeoff, generally in the early morning hours to take advantage of calm wind conditions and optimum sun angles for the sensor. Flexibility in the schedule allowed the

[67] Philip Hall, Brent Cobleigh, Greg Buoni, and Kathleen Howell, "Operational Experience with Long Duration Wildfire Mapping."

flight to slip to Friday in the event of unforeseen difficulties, creating a practical limitation of one flight per week maximum.[67]

Loss of command link and emergency procedures

Increasing use of unmanned vehicles in the national airspace has raised concerns about their potential to cause harm to persons and property in the air and on the ground. The Ikhana WSFM team worked hard to reduce potential risks to an acceptable level. Mission planners must allow for contingencies

With its sensor pod under the left wing, NASA's remotely piloted Ikhana unmanned aircraft cruises over California during the Western States Fire Mission.
NASA

and alternatives for every point along the flight track. Due to a lack of real-time situational awareness, all available emergency landing sites are identified prior to each mission.

During each flight, a lost-link mission plan is loaded and maintained in the aircraft's computer in the event of a total communications failure between the Ikhana and the GCS. In the event of a command-and-control link malfunction, the aircraft will autonomously proceed to a specified point over unpopulated terrain and eventually return to Edwards and a location that virtually guarantees pilots will be able to regain control of the aircraft. The pilot continually updates the lost-link mission plan during the flight to reflect the current aircraft position.

The Ikhana's electric power is normally generated by engine operation. In the event of electrical system failure, the vehicle has batteries capable of supplying power for approximately three hours. This would allow the aircraft to fly approximately 400 nmi, including descent and landing maneuvers. Alternate contingency landing sites were identified and made available for instances in which Ikhana was more than

400 nmi from Edwards when a malfunction occurred. For the Western States Fire Mission, primary emergency landing sites included Mountain Home Air Force Base, Idaho and Michael Army Airfield, Utah. Agreements with the Air Force and Army specified risks and hazards associated with landing the Ikhana under satellite control as well as necessary ground procedures following landing at remote locations.

If the aircraft loses thrust – due to engine or propeller malfunction, or any other reason – an emergency landing site must be identified within the Ikhana's gliding distance based on mission altitude. At a cruising altitude of 23,000 feet, the aircraft can glide for approximately 50 nmi. Planners determined secondary emergency landing sites spaced no more than 100 nmi apart throughout the entire area defined by each COA. Additionally, a runway of at least 4,000 feet in length, and preferably 5,000, is necessary for a safe landing.

Because Ikhana lacks sufficient capability to detect and avoid other aircraft in the airfield's landing pattern or in the nearby vicinity, the FAA directed that these sites could not be active civil or joint-use

NASA research pilot Mark Pestana prepares to fly the Ikhana unmanned aircraft remotely from the GCS at NASA Dryden.
NASA

airports. NASA Dryden officials stipulated that no military airports could be considered without prior coordination. As best it could, a five-member team from NASA Dryden spent two months surveying the Western United States for potential landing sites, reviewing those sites with the Dryden Range Safety Office, and gathering additional information on sites deemed acceptable.[68]

Suitable landing sites included dry lakebeds, abandoned runways, farm fields, and other open spaces that would serve as suitable landing sites away from populated areas. Flight personnel rated each site according to its suitability for a safe landing. Marginally suitable sites would only allow for a crash landing that would not endanger the public. This analysis resulted in compilation of a database of more than 280 emergency landing sites, many of which would be appropriate for any given flight. During each flight, the Mission Director would maintain real-time situational awareness of the nearest suitable emergency landing site.[69]

Ikhana project engineer Kathleen Howell and project manager Brent Cobleigh check the planned flight paths in Ikhana's ground control station before takeoff.
NASA

[68] Gregory P. Buoni and Kathleen M. Howell, "Large Unmanned Aircraft System Operations in the National Airspace System."

[69] Philip Hall, Brent Cobleigh, Greg Buoni, and Kathleen Howell, "Operational Experience with Long Duration Wildfire Mapping."

An additional challenge in making an emergency landing is that real-time control of the aircraft via the Ku-band satellite link has an inherent signal delay of up to two seconds. During the landing phase, when pilot control inputs are particularly dynamic and need to be applied in a timely manner, this delay can lead to pilot-induced oscillation. The pilot must anticipate the control inputs necessary to avoid making an incomplete flare or hard landing, or even loss of aircraft control. In order to reduce risk and satisfy both FAA and NASA management's acceptance of this risk, NASA pilots spent many hours flying the Ikhana simulator. Hosted on a desktop computer system, the simulator is capable of emulating the signal delay. The pilots learned to compensate by reducing power and flaring early – "flying in the future," as Pestana put it.[70]

Ready to fly

The WSFM team's desired COA area included all parts of the Western United States that have shown a history of wildfires. This includes a region from the Canadian border to the Mexican border and from the Pacific Ocean to Denver. Planners made every effort to include as much of this region as possible for the 2007 fire season.

In late July 2007, NASA received approval for initial WSFM flights but with significant restrictions. The COA as approved by the FAA was not identical to that requested in the application. The most notable change was a limitation of the operating boundary to within 75 nmi from three defined backbone routes (one in each zone), thus preventing the AMS sensor from imaging any fires beyond this limit. The backbone had no spokes; unfortunately backbone routes had not been selected so as to pass over areas with a high-likelihood of wildfires.

Flights were also prohibited in areas that might be affected by known solar storms, planned Global Positioning System (GPS) testing, or outages in Receiver Autonomous Integrity Monitoring (a system that indicates the integrity of GPS signals). Weather restrictions included a prohibition of flights into turbulence or icing conditions forecast to be moderate to severe. Despite these limitations, the Ikhana team was ready to perform operational missions.[71]

[70] Mark Pestana, comments on draft manuscript, December 2008.

[71] Philip Hall, Brent Cobleigh, Greg Buoni, and Kathleen Howell, "Operational Experience with Long Duration Wildfire."

Chapter Three
THE FIRE DOWN BELOW

The first four Ikhana flights in the national airspace set a benchmark for establishing criteria for future science operations. During these missions, the Ikhana traversed eight western U.S. states, collecting critical fire information and relaying data in near real-time to fire incident command teams on the ground as well as to the National Interagency Fire Center (NIFC) in Boise, Idaho. Data from the AMS-

to see and use data in as little as 10 minutes after it was collected.[72]

The Google Earth DSS CDE also supplied other real-time fire-related information including satellite weather data, satellite-based MODIS fire data, Remote Automated Weather Station readings, real-time lightning strike detection data, and other critical fire database source information. Google Earth

Carrying its sensor pod, the Ikhana banks away during a checkout flight in the Western States Fire Mission.
NASA

Wildfire instrument was downlinked to the GCS and then transferred to a server at NASA-Ames and autonomously redistributed to a Google Earth data visualization capability – CDE – that served as a Decision Support System (DSS) for fire data integration and information sharing. This system allowed users

imagery layers allowed users to see the locations of manmade structures and population centers in the same display as the fire information. Shareable data and information layers, combined into the CDE, al-

[71] Philip Hall, Brent Cobleigh, Greg Buoni, and Kathleen Howell, "Operational Experience with Long Duration Wildfire."

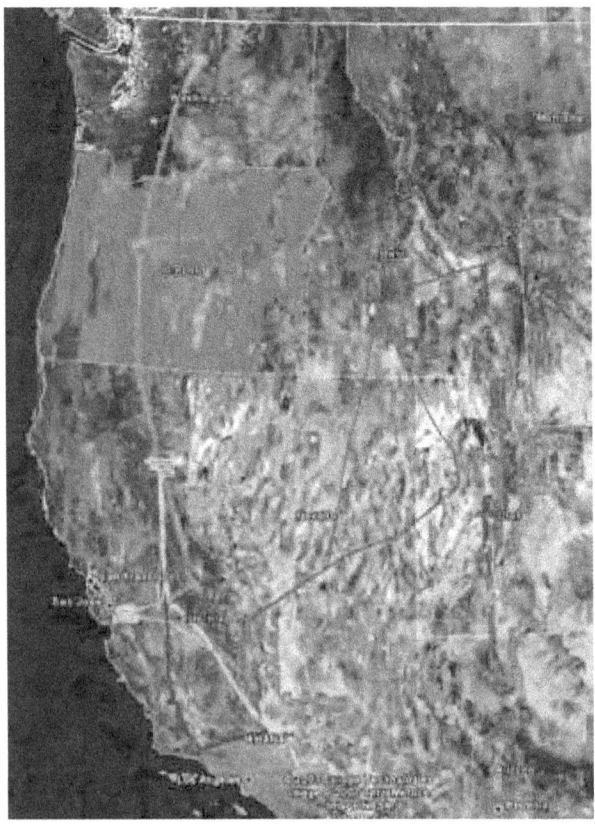

Flight tracks in red, green, blue, and yellow indicate the routes of the first four Ikhana Western States Fire Missions.
NASA/Google

lowed the incident commanders and others to make real-time fire management strategy decisions. Personnel throughout the U.S. who were involved in the mission and imaging efforts also accessed the CDE data. Fire incident commanders used the thermal imagery to develop management strategies, redeploy resources, and direct operations to critical areas such as neighborhoods.[73]

During these four sorties, the Ikhana flew a total of 56 hours, operating in the national airspace for nearly 50 hours. The aircraft overflew and imaged a total of 26 fires, providing real-time data to incident commanders and the NIFC.[74]

The Ikhana team also collected post-fire burn-assessment imagery over various fires to aid teams in fire ecosystem rehabilitation on those major events. The Burned Area Emergency Rehabilitation

(BAER) project examined the effects of the fire on soil, watershed, wildlife, vegetation, and other resources for potential threats to life and property. After experts complete an assessment of an area, the BAER team can propose alternative prescriptions for a rehabilitation plan to local Forest Service officials to minimize the fires' effects on resources, life, and property within and directly adjacent to the fire. Rehabilitation efforts can include such activities as seeding, use of hay mulch, and vegetation planting designed specifically to meet resource objectives, such as minimizing erosion.[75]

Three additional missions were flown after requests from several government agencies responding to a wildfire emergency in Southern California. These flights demonstrated the Ikhana team's ability to provide critical sensor data and distribute it to users in a timely manner under extremely challenging circumstances.[76]

Four states ablaze

The first WSFM flight took place Aug. 16 and was limited to about 10 hours duration in order to verify planning and coordination procedures. During an unexpectedly short pre-mission teleconference, representatives of the FAA ARTCCs accepted route plans submitted by the NASA team even though one wildfire was at the 75 nmi limit from the Route A backbone. The FAA granted permission for the Ikhana to fly slightly beyond the COA boundary to cover this fire.[77]

Based on the approved route, approximately 12 emergency landing sites were selected from a list of more than 280 pre-designated sites within the flight zones. During the mission it appeared that local air traffic controllers had, for the most part, been briefed in advance and were expecting the aircraft. Over

[72] "Completed Missions," Wildfire Research and Applications Partnership (WRAP), http://geo.arc.nasa.gov/sge/WRAP/current/com_missions.html, 2008, accessed June 10, 2009.

[73] Ibid.

[74] Ibid.

[75] "BAER: Burned Area Emergency Rehabilitation," Lytle Fire 2003, IncidentControl.com, http://www.incidentcontrol.com/lytlefire/b_a_e_r.htm, October 2003, accessed June 10, 2009.

[76] Gregory P. Buoni and Kathleen M. Howell, "Large Unmanned Aircraft System Operations in the National Airspace System."

[77] Ibid.

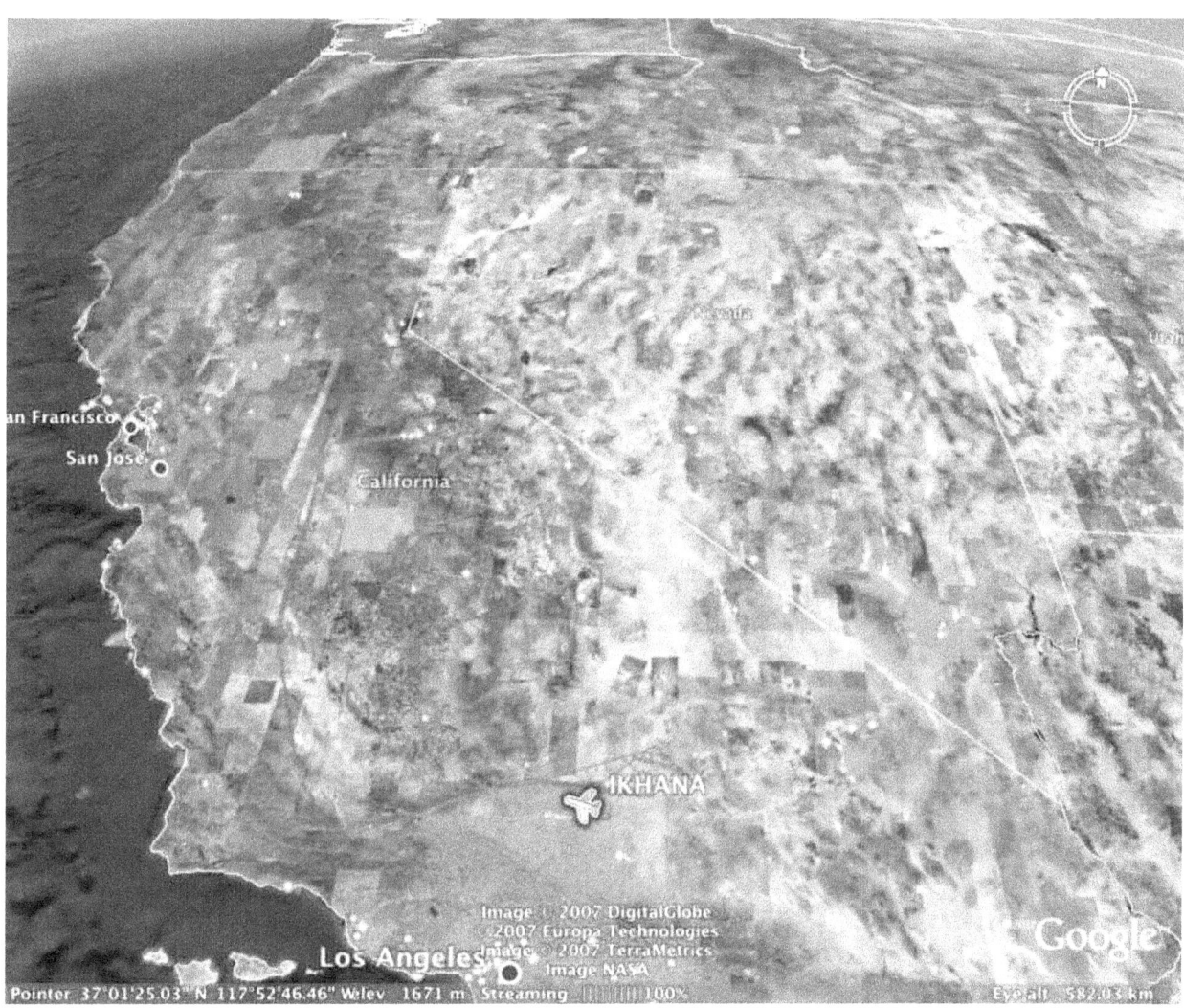

During WSFM-1, sensors on board the Ikhana covered the Zaca fire (lower left), Tar fire, Babcock fire, and Colby fire (upper left). NASA/Google

the course of 9.5 hours, the Ikhana flew over four wildfires in southern and central California (the Zaca, Tar, Colby, and Babcock fires), covering a distance of 1,400 nmi. Incident commanders at the Zaca fire were able to access and use data between four and nine minutes after acquisition. The Zaca fire incident commander was so impressed by the WSFM team's contributions that he appealed to California Senator Barbara Boxer's office for additional Ikhana imaging flights.[78]

During WSFM-2 on Aug. 30, the Ikhana remained aloft for 16.1 hours. Data passes were accomplished over five active fires and several previous burn areas, in California (Jackrabbit, Zaca, Tar, Colby, and Babcock), Nevada, Utah, Idaho (Trapper Ridge and

Castle Rock), Montana (WH Complex fires), and Wyoming (Columbine, Hardscrabble, and Granite Creek). The total distance covered was about 2,500 nmi. The WSFM team had to cope with momentary loss of color nose-camera imagery during takeoff as well as a mission-planning software editing problem after takeoff and before landing.[79]

Researchers and firefighters were extremely interested in getting coverage of wildfires north of the approved COA boundary in Idaho and Montana, so the Ikhana team requested a COA extension from the FAA. Unfortunately, the request was denied. Since multiple large wildfires were burning within the approved COA boundaries in Idaho, Montana, and Wyoming, the team submitted a second route

[78] Ibid.

[79] Ibid.

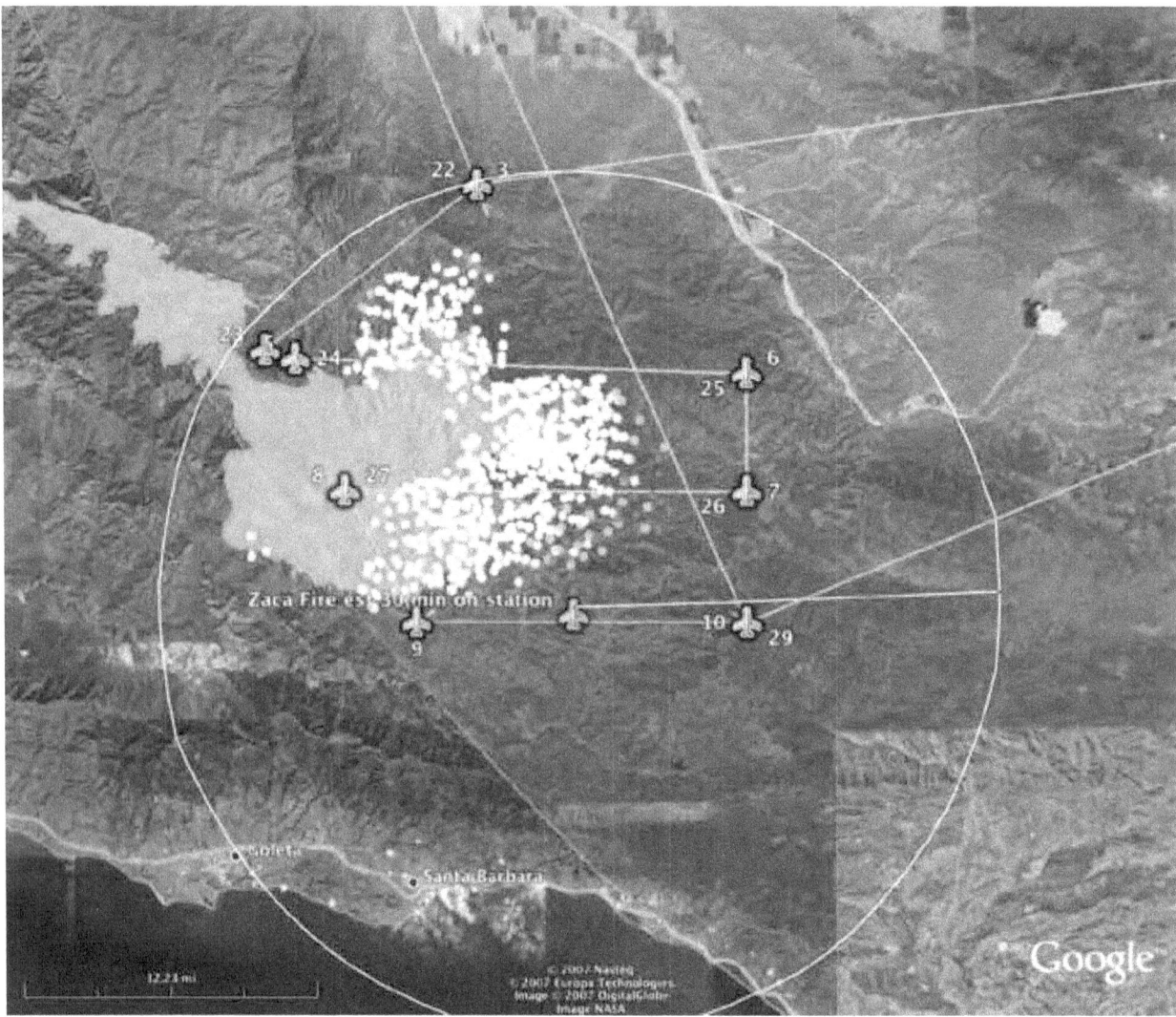

In the WSFM-1 flight plan, the known perimeter of the Zaca fire is indicated in light green. MODIS hot detects are shown in yellow, orange, and red.
NASA/Google

plan using the original COA boundary but the real challenges were just beginning.

In the wake of a recently created Air Traffic Control Assigned Airspace (ATCAA – extension of Military Operating Area activities above 18,000 feet), officials at Salt Lake Center requested several waypoints along the Ikhana's planned route be moved so the aircraft could avoid this area. The suggested alternative route would have passed directly over one of the keep-out zones and was unacceptable.

Eventually, a route was approved that successfully avoided both the ATCAA and the populated keep-out zone. Now, however, there was a conflict with GPS testing that impacted part of the route and the mission was delayed 24 hours. Because of the

areas covered, the Ikhana team always had to coordinate flights so as to avoid conflict with GPS testing activities at Naval Air Warfare Center China Lake, the Nevada Test and Training Range, and the Utah Test and Training Range.

On the morning of the rescheduled flight, the Edwards runway was closed for repairs, resulting in an additional postponement. The mission was flown a week later, after coordinating with scheduled GPS tests.

Two data runs on the Castle Rock fire were coincident with overpasses by a satellite carrying a MODIS sensor. Earth Observing System spacecraft – called Terra and Aqua – equipped with MODIS sensors view the entire Earth's surface every one to two

A mosaic of images of the Zaca fire was georectified and draped over Google satellite imagery of the terrain. Bright areas indicate active fires.
NASA/Google

days, acquiring data in 36 spectral bands to improve researchers' understanding of global dynamics and processes occurring on the land, in the oceans, and in the lower atmosphere. Such data can be compared with that collected with the Ikhana's more sensitive instruments.[80]

This mission provided an opportunity to demonstrate real-time weather re-routing. En route to Utah, FAA officials allowed a significant deviation around thunderstorms in northern Nevada, demonstrating that they could treat the Ikhana like any other aircraft in the national airspace.

The third sortie followed an ambitious flight plan for the first true long-duration mission. FAA officials were initially concerned about plans to fly the Ikhana in close proximity to airspace that was used heavily during daylight hours. After considering alternative routes, the time of day Ikhana was to fly though the airspace in question, and assurances that the aircraft could be safely re-routed in real-time, FAA officials approved the route as originally submitted.

GPS testing delayed the planned takeoff time,

[80] Ibid.

A close-up image of the southeastern portion of the Zaca fire reveals hot spots in rugged terrain.
NASA/Google

resulting in changes to the timing of passes over wildfires along the route. This necessitated reversal of the route's direction of flight in order to take advantage of more favorable sun angles over the active fires and burn areas to be imaged. Fortunately, the FAA permitted this change.[81]

WSFM-3 covered 12 fires in California (Butler, North, Fairmont, Grouse, Zaca, Bald, Moonlight, Lick), Oregon (GW and Big Basin), and Washington (Domke Lake and South Omak) in the course of a 20-hour mission on Sept. 7 and 8. During the 3,200 nmi flight, the Ikhana flew from Southern California to within 50 miles of the Canadian border. Multiple passes were made over four of the major fires and several fires were imaged during night and day passes. Five imaging lines were run over the Zaca fire burn area using a spectral band suitable for the BAER project. Ikhana WSFM project representatives were deployed to the NIFC to assist with data evaluation. Teams were also sent to two active fires to help incident commanders with data interpretation and CDE visualization. One team, led by Tom Zajkowski of the U.S. Forest Service, assisted with the GW fire in Oregon, while another from NASA Ames worked the

[81] Ibid.

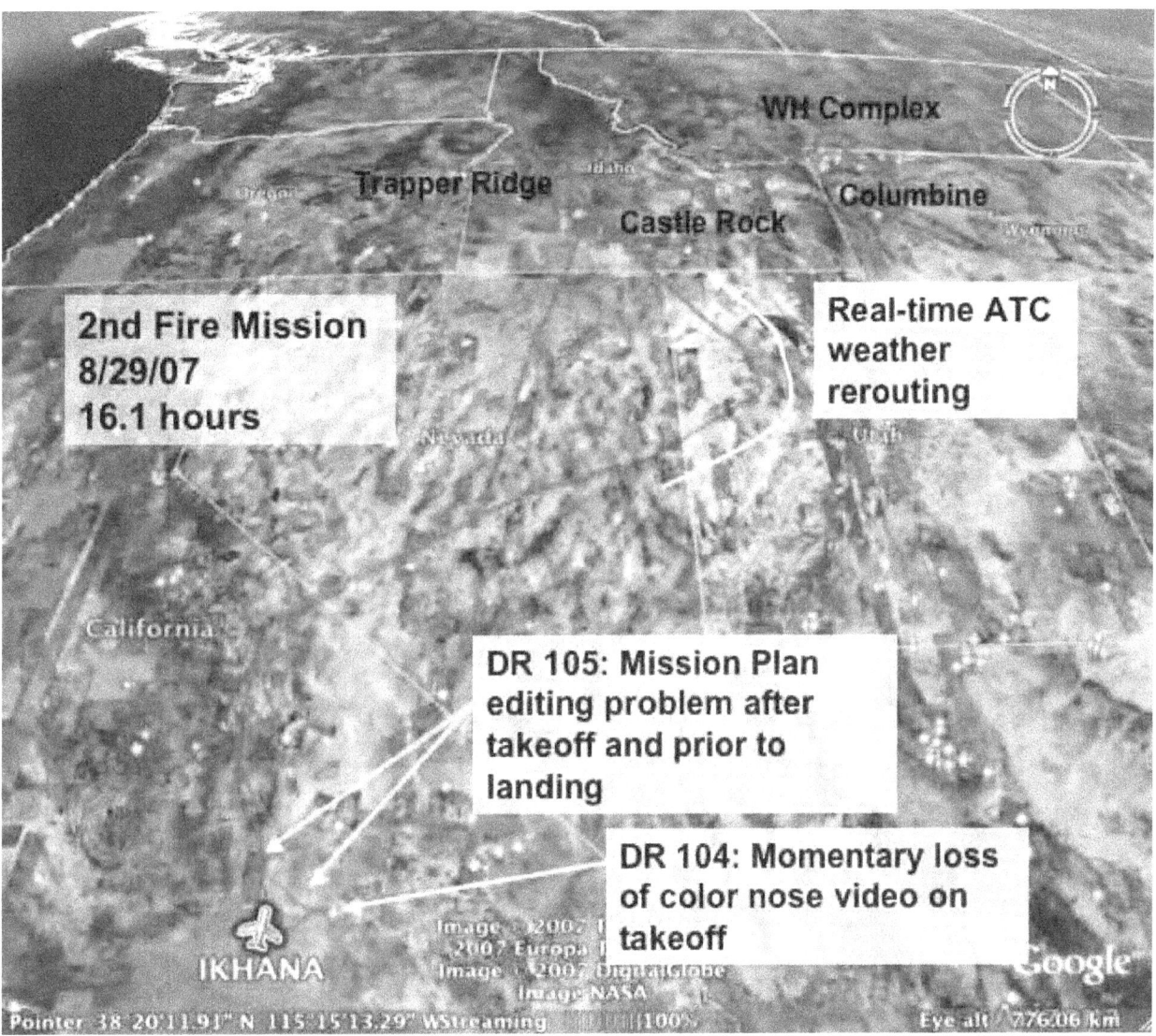

The route for WSFM-2 covered fires as far away as Idaho, Montana, and Wyoming. Air traffic controllers rerouted the Ikhana to avoid potential weather hazards over northern Nevada.
NASA/Google

Lick fire near San Jose, Calif.[82]

FAA Air Traffic Control flexibility was demonstrated during a real-time movement of one of the wildfire incident loiter locations. Coordination with controllers allowed a seven-mile deviation from the filed flight plan at the Lick fire, which had seen significant movement since the route plan was developed and submitted three days prior to flight. The Ikhana team had excellent FAA support throughout the mission. Mission planners opted to bypass a repeat collection opportunity on the Domke Lake Fire in

Washington due to flight time allocation restrictions. Additionally, the Los Angeles Air Traffic Control Center could not give sufficient routing to image the Butler fire due to early evening air traffic in the vicinity of the Ikhana's flight path.[83]

Near the end of the flight, the aircraft exhibited a pitch disturbance when the pilot keyed his microphone. Discussions with Air Force and General Atomics personnel revealed that this same behavior had been seen on several early production aircraft during the previous few weeks. Electromagnetic interference between wiring bundles was suspected.

---

[82] Brent Cobleigh, Ikhana Flight Reports, NASA Dryden Flight Research Center, Edwards, CA, 2007.

[83] Ibid.

The Castle Rock fire burned hillsides above populated areas. Hot spots are shown in red.
NASA/Google

At left, MODIS data (red and orange dots) from sensors on board the EOS Terra satellite were compared with Ikhana AMS hot detects (yellow). Data collected with the Ikhana were substantially more detailed.

The fourth mission, on Sept. 27, focused on BAER data collection with overflights of four burn areas throughout California (Butler, Grouse, Lick, and Moonlight). Because this mission primarily involved returning to areas covered during earlier flights, the FAA approved the three-day-prior flight plan submission with little discussion. The Ikhana's route covered 1,800 nmi in 9.9 hours. During the final hour of the flight, the aircraft climbed to 40,000 feet with approximately half its total fuel load to collect fuel flow performance data with the pod attached.[84]

[84] Ibid.

During WSFM-3 the Ikhana covered 12 fires in California, Oregon, and Washington. The mission was highly successful but a few instances of uncommanded pitch bobble were cause for concern. NASA/Google

In this 3-D image of the Lick fire zone, shades of red and purple highlight the burn area. Yellow hot spots indicate active fires. NASA/Google

Use of 3-D projection, as in this image of the Moonlight fire, allowed incident commanders to better visualize challenging terrain while developing plans to deploy firefighting resources.
NASA/Google

Southern California emergency firestorm support

In late October 2007, hot, dry Santa Ana winds fanned nearly a dozen major wildfires in the Los Angeles and San Diego regions of Southern California. Spread by 40-80 knot winds, flames pushed through populated areas and devastated vast regions of wilderness, neighborhoods, and businesses. More than 500,000 people had to be evacuated in the endangered areas. On Oct. 22, officials at the California Governor's Office of Emergency Services and NIFC requested NASA airborne remote sensing resources to assist in monitoring the fast-moving fires. The Wildfire Research and Applications Partnership team at Ames and the Airborne Sciences team at Dryden accepted the challenge to rapidly develop a series of mission profiles to support emergency services personnel battling the firestorm. Although the sensor pod had been removed from the Ikhana following the recently completed WSFM flights and the aircraft was being configured to support the Fiber Optic Wing Shape Sensing project, the two NASA teams quickly reconfigured the platform and sensor and prepared

A helicopter drops water and retardant on the Harris fire near the Mexican border during the 2007 Southern California wildfire emergency. FEMA/Andrea Booher

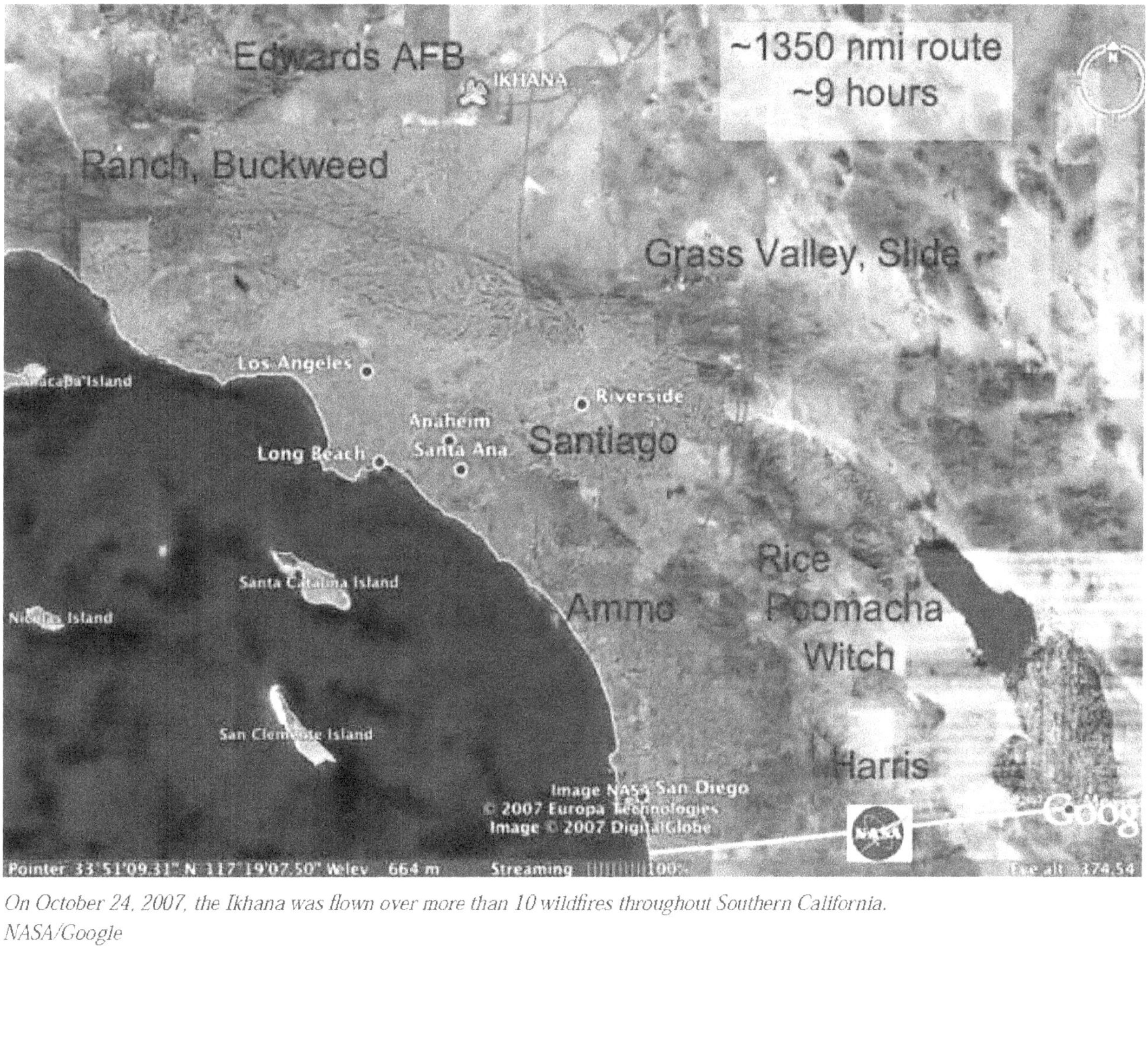

On October 24, 2007, the Ikhana was flown over more than 10 wildfires throughout Southern California. NASA/Google

In order to collect this mosaic of images of the Harris fire, the Ikhana had to be flown within 10 miles of the Mexican border. Active fire is indicated in yellow.
NASA/Google

for an emergency support mission just two days after receiving the request from the governor's office.

The Ikhana team experienced a flurry of activity while striving to monitor fires from the vicinity of Lake Arrowhead in the San Gabriel Mountains to as far south as San Diego County. Because the area of interest was south of the approved COA boundary, the FAA quickly amended the existing COA to authorize the Ikhana to fly to within 10 nmi of the U.S.-Mexico border. The three-day advance flight plan submission request deadline was reduced to 48 hours, and later to 24 hours in order to expedite the mission planning process.

At one point, planners at Dryden learned that GPS testing in the vicinity of Nellis Air Force Base, NV, would impact the mission. Consultation between NASA and Air Force officials ensued regarding the emergency

situation. Nellis officials agreed to a three-day stand-down of GPS testing so Ikhana missions could proceed. The FAA was notified of the coordination effort.[85]

On the morning of Oct. 24, the Ikhana took off to begin a nine-hour mission over more than 10 California wildfire locations. The airplane covered a distance of 1,350 nmi while collecting thermal imagery data over the Harris, McCoy, Witch/Poomacha, Coronado Hills, Rosa, Slide, Grass Valley, Buckweed, Ranch, Magic, and San Clemente fire zones. Real-time mission re-planning allowed adaptation to changing fire conditions. This capability allowed coverage of a fire that broke out within the boundaries of the U.S. Marine Corps base at Camp Pendleton.

[85] Gregory P. Buoni and Kathleen M. Howell, "Large Unmanned Aircraft System Operations in the National Airspace System."

This 3-D image shows the Harris fire tracing rugged mountain ridges. Blue and purple indicate charred terrain, with active fire in yellow
NASA/Google

The information collected with the Ikhana was provided to fire Incident Command Centers (ICC) and various County Emergency Operations Centers (EOC) as quickly as possible. By the time the Ikhana landed in the late afternoon the sensor had collected, processed, and transmitted approximately 100 thermal-infrared data scenes and numerous fire-detection shape files. These products were in the hands of disaster managers within minutes of collection. This success led to a request for further support by the agencies and an additional three missions were subsequently flown.[86]

WSFM-6, a 7.8-hour sortie on October 25, was another emergency wildfire response mission at the request of the California Office of Emergency Services

(CA-OES). All active fires were visited twice during the mission, including the Harris, Rice, Witch/Poomacha, Ammo, Slide, Grass Valley, Buckweed, and Ranch fires along a 1,350 nmi route. Imagery was fed to officials at the Federal Emergency Management Agency, NorthComm fire dispatching agency in San Diego County, CA-OES, and on-scene fire commanders. A post-flight debrief with FAA officials – required after each flight – confirmed that no air traffic control issues were encountered during the mission. General Atomics provided two pilots during the mission to provide breaks to the two NASA pilots. NIFC officials requested that the project team continue flying the Ikhana the following day.[87]

[86] "Completed Missions," Wildfire Research and Applications Partnership.

[87] Gregory P. Buoni and Kathleen M. Howell, "Large Unmanned Aircraft System Operations in the National Airspace System."

In a 3-D mosaic of images of the Grass Valley and Slide fire areas, active fire is indicated in yellow while hot, previously burned areas appear in shades of dark red and purple. Unburned areas are green and brown.
NASA/Google

WSFM-7 took place on Oct. 26. For the third consecutive day, the joint NASA/Forest Service team responded to the California wildfire emergency at the request of the NIFC and CA-OES. The NASA Dryden Range Safety Office (RSO) approved a route that included all the fires previously imaged as well as the Santiago fire in Orange County. With dense population zones on three sides, the RSO had to identify safe entry and exit routes to the burn area. Because so many people had been evacuated from threatened areas, one of the previously existing keep-out zones was removed, allowing expanded imaging of the Poomacha fire.[88]

During an early morning preflight briefing, planners decided to reverse the direction of the route due to

weather concerns. Since Air Force weather personnel at Edwards were not yet on duty to accept this flight plan revision, the Ikhana team filed the flight plan directly with a civilian Flight Service Station (FSS). This posed a challenge because the plan began with the aircraft exiting a restricted area and ended with it entering a restricted area, an unusual circumstance not accounted for in standard FSS flight plan procedures. Personnel at the FSS resolved the issue by entering an explanation in the "Remarks" section of the flight-plan form. A second challenge arose when planners had to explain to the FSS that there were "0" souls onboard since the Ikhana carried no pilot or passengers.

General Atomics provided one relief pilot during the 8.7-hour mission that traversed 1,350 nmi. FEMA

[88] Ibid.

Crews man a firebreak on the Poomacha fire in brush-covered hills west of the Salton Sea. Incident commanders studied AMS imagery from the Ikhana while developing a deployment strategy for firefighters combating the blaze.
FEMA/Andrea Booher

Expanded imagery of the Santiago fire was possible because a large number of people had been evacuated from threatened areas.
NASA/Google

BAER imagery shows the extensive burn area caused by the Santiago fire. Note the proximity of neighborhoods on the southern perimeter.
NASA/Google

officials again requested that Ikhana flights continue through the weekend. A sortie was scheduled for Oct. 27 but was cancelled due to fog over the desired coverage areas. The AMS sensor cannot penetrate fog. The final mission of the 2007 fire season, WSFM-8, was completed Oct. 28. It encompassed 11 fires and 1,350 nmi in just 7.1 hours, including a BAER assessment of the one-year-old Esperanza fire zone.

Active fires included Harris, Rice, Witch, Poomacha, Ammo, Santiago, Slide, and Grass Valley. BAER imagery was also taken at the Buckweed and Ranch fire zones. Once again, General Atomics provided a relief pilot.[89]

Department of Defense officials requested authorization to fly other unmanned aircraft over the fire areas,

[89] Cobleigh, Ikhana Flight Reports, NASA Dryden Flight Research Center.

Real-time mission re-planning allowed adaptation to changing fire conditions. This capability allowed coverage of the Ammo fire that broke out within the boundaries of the U.S. Marine Corps base at Camp Pendleton.
NASA/Google

BAER imagery shows the extent of the Ammo fire burn area.
NASA/Google

but the FAA limited the number of unmanned aircraft in the Los Angeles Center's airspace to one at a time, and the Ikhana mission was given priority. Ikhana project management learned that the missions would be briefed in the White House Situation Room.[90]

In responding to the California wildfire emergency, the Ikhana team faced a significant challenge. Each mission profile changed depending on the emergency priority for each fire as conditions changed and as data were relayed in real-time to the ICCs and EOCs. Ikhana's sensor collected approximately 100 images each day and relayed them to the ICC and EOC personnel. As fire activity decreased, some of the mission focus shifted to collection of BAER imagery to support ecological recovery efforts in the affected areas. Researchers used the multi-channel capabilities of the AMS-Wildfire sensor to execute real-time sensor reconfiguration during the course of the mission for either active fire mapping or burn-severity data collection.

During the four Southern California wildfire-imaging missions, Ikhana UAS operators logged a total of approximately 36 flight hours and collected more than 400 AMS-Wildfire images. Data was used tactically by the individual ICCs as well as strategically by the EOC to allocate firefighting resources arriving

During the four Southern California wildfire emergency missions, Ikhana operators logged more than 36 flight hours and collected more than 400 images with the AMS sensor package.
NASA/Google

from other states to fires deemed highest priority based on analysis of AMS imagery. Incident response teams readily adapted Google Earth visualization capabilities into their operations to allow AMS-Wildfire information to be integrated with other critical data layers such as weather, terrain, and local population density. Total downloads of Ikhana imagery exceeded 40,000 files. A spectacular success, the emergency support

90 Gregory P. Buoni, "Ikhana Weekly Notes," NASA Dryden Flight Research Center, Edwards, CA, Oct. 26, 2008.

missions demonstrated the importance of integrating capabilities of various federal agencies for improved disaster response.[91]

WSFM principal investigator Vincent Ambrosia had anticipated an event like the wildfire siege in Southern California occurring in October. Plans were already in place to have team members at various fire camps to assist with integration of data and imagery derived from the AMS-Wildfire sensor on Ikhana while other personnel remained in place at Dryden, Ames, Google, and the National Interagency Fire Center.

"When the call came on Monday from the National Interagency Fire Center, the California Governor's Office of Emergency Services, and colleagues within the Incident Command structure on the fires, we were ready to quickly deploy our teams and initiate a mission plan to over fly the fires and provide critical thermal infrared intelligence on the various wildfires," Ambrosia explained.[92]

2008 California fire missions

Firestorms again swept California in early summer 2008, bringing calls from the state capitol for more

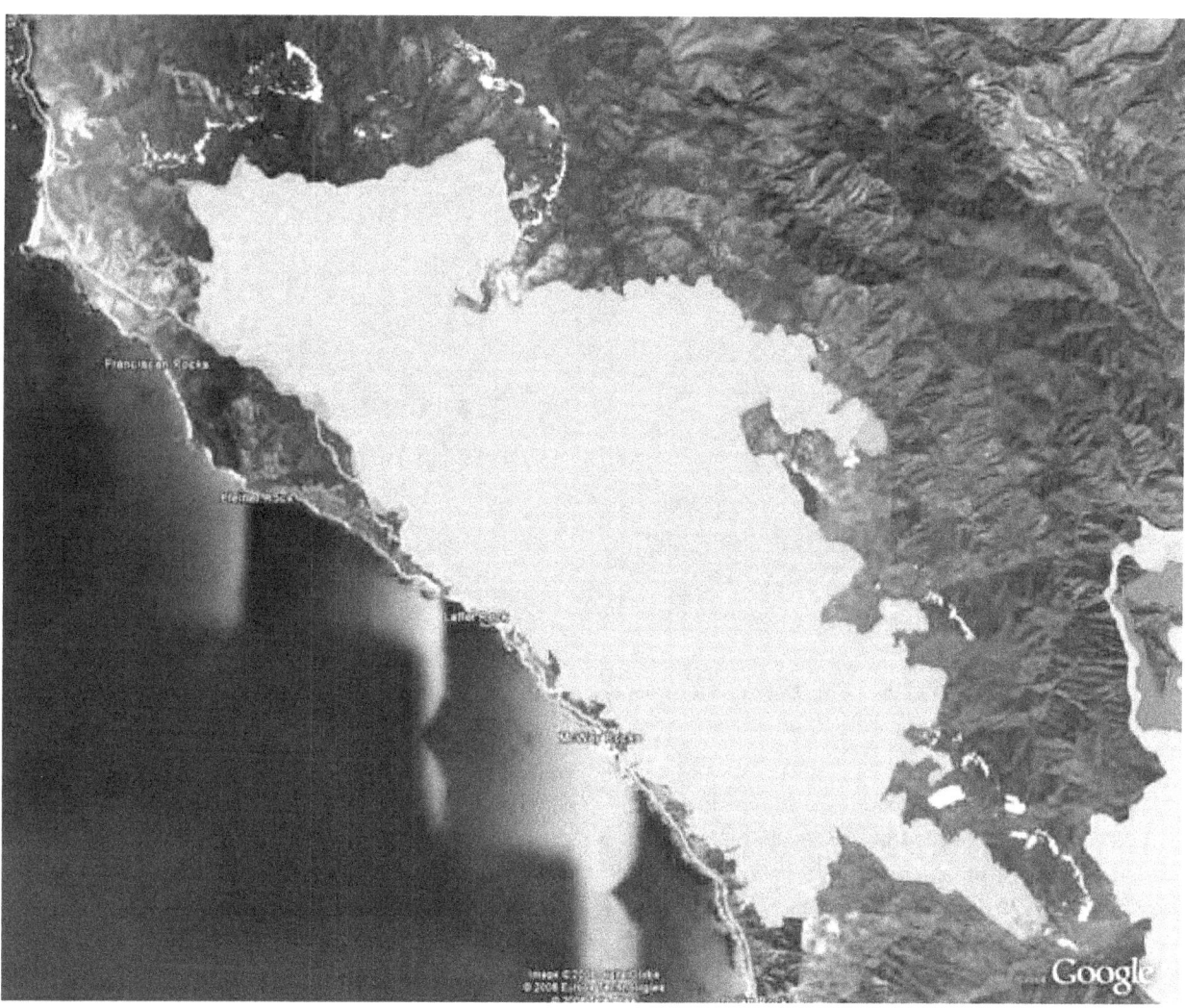

Scientists used AMS imagery to map actively burning areas of the Basin Complex fire in Monterey County, Calif., on July 8, 2008, as seen here in yellow. Known fire perimeters are seen in green, purple, blues and reds. Multiple California state agencies distributed the information to firefighting officials within minutes of collection, enabling near-real time tactical decision-making. NASA/Google

[91] "Completed Missions," Wildfire Research and Applications Partnership.

[92] Aviation.com staff, "NASA Flies Ikhana UAV to Help California Firefighters," Aviation.com, http://www.aviation.com/technology/071024-nasa-ikhana-california-wildfires.html October 2007, accessed June 10, 2009.

The AMS captured this image of the Gap fire in Santa Barbara County, Calif., on July 8, 2008. The yellow areas depict actively burning fires. The red, pink, and blue tones map burned areas. Multiple California state agencies distributed the information to fire officials minutes after collection for analysis of new fire locations and fire size.
NASA/Google

Ikhana missions. The NASA/Forest Service team gathered six weeks earlier than originally planned and started working on plans to provide state and federal agencies with critical fire intelligence.

"Because Forest Service assets are stretched thin, NASA was asked to provide additional resources as a supplement to existing infrared fire imaging operations," said Jim Brass, co-principal investigator for the WSFM at Ames.[93]

"California's unprecedented number of fires this early in the season make it all the more important that we use every tool at our disposal to protect property and save lives, " said Gov. Schwarzenegger. "NASA's Ikhana is one more incredible tool that we are able to use this year to bring real-time pictures and data to fire commanders, even when our other aircraft are unable to fly."[94]

The WSFM team applied for a new COA. It was

[93] Beth Hagenauer and Mike Mewhinney, "NASA Responds to California Wildfire Emergency Imaging Request," Release 08-30, NASA Dryden Flight Research Center, Edwards, CA, July 11, 2008.

[94] General Atomics Aeronautical Systems press release, "Governor Schwarzenegger and NASA Highlight Infrared Scanning Technology Helping to Fight California's Wildfires" (GAAS:525:08), July 14, 2008.

A flight plan for the second WSFM sortie of 2008 shows the aircraft's route over the Sierra Nevada mountain range. Keep-out areas are indicated in yellow and red.
NASA/Google

approved without the limitation of having to fly within 75 nmi of the backbone route. There was, however, a restriction to remain within 50 nmi of restricted airspace and Military Operating Areas.

The first attempted WSFM sortie of 2008 began on the morning of July 1. Planners scheduled passes over 12 blazes spanning California from the southern Sierra Nevada mountain range to Big Sur on the northern California coast. The plan included coverage of the Havilah, Clover, Piute, Basin Complex, Friant, North Mountain, American River Complex, Yuba River

Complex, Amedee, Cub Complex, Antelope, Gap, and Mill Complex fires. Just 1.5 hours into the flight, following a pass over the Piute fire, a problem with the sensor caused the mission to be aborted. Technicians added a heating system to the sensor prior to the next flight attempt.[95]

The Ikhana team re-flew the sortie on July 8, im-

[95] Ikhana flight plans (6/29/08) and meeting notes, Ikhana Team Meeting, NASA Dryden Flight Research Center, Edwards, CA, July 3, 2008.

Flight plan detail shows planned routes over the Cub Complex and Canyon Complex fires.
NASA/Google

aging almost 4,000 square miles from Santa Barbara north to the Oregon border. This time, everything went as planned and the aircraft remained aloft for 9.9 hours while covering 10 individual and complex fires along a route that took the aircraft west over the Sierra Nevada range, north to the Cub Complex fires, and south to the Basin fire in Monterey County and the Gap fire in Santa Barbara County. Imagery was transmitted to the Multi-Agency Coordination Center in Redding, Calif., and the State Operations Center in Sacramento for distribution to incident commanders in the field. Data from the AMS-Wildfire sensor was used to identify previously undetected hotspots near the town of Paradise in Butte County. As a result, some 9,500 residents were evacuated and firefighting resources were re-allocated to protect populated areas.[96]

The second mission flew as scheduled on July 19. In a little over five hours, Ikhana and the AMS-Wildfire

sensor covered the American River Complex, Camp, and Canyon Complex fires in central California. The Ikhana team also took the opportunity to image the results of a mudslide that destroyed several structures near Independence, Calif., and to briefly revisit the Piute fire area. During the mission, the team received feedback from Canyon Complex Situation Unit Leader Randy Herrin.

"Thanks for the imagery on the Canyon Complex," he wrote in an email to Ikhana project manager Tom Rigney. "I was able to follow along on the CDE and video and show the project to the Operations Chief and Deputy IC. They were impressed, to say the least. The imagery showed a significant amount of heat in the SW of our complex, which we were not expecting, so that was good to know. Congratulations to everyone on another successful mission."[97]

In a separate message to Ikhana scientist Steve Wegener, Herrin wrote, "The products that we've received here on the incident are very useful and well received. There were several pairs of eyes opened up here to what the future may hold, and how close that future may be."[98]

The final WSFM sortie of 2008 took place on Sept. 19 with planned coverage over the Rattle and Bear Wallow fires in Oregon and the Klamath Theater, Cascadel, and Hidden fires in central California. Due to predicted cloud cover over the two Oregon fires, they were eliminated from the mission data acquisition plan. Focus, therefore, was on the central California fires.

Preparations began at NASA Dryden in the darkness of the predawn hours. Mark Pestana and Herman Posada completed preflight activities in the GCS for an 8 a.m. takeoff. Forty-five minutes later Posada, using the forward-facing camera, spotted two plumes of gray smoke rising above a mountain ridge. Pestana called Oakland Center to request a 60-minute loiter time over the fire area.

At the request of the science staff, the Ikhana was flown first over the Hidden fire, followed by two passes

[96] General Atomics Aeronautical Systems press release, "Governor Schwarzenegger and NASA Highlight Infrared Scanning Technology Helping to Fight California's Wildfires." Interview with Thomas Rigney, Aug. 13, 2008.

[97] Email from Thomas K. Rigney, Ikhana Project Manager, to distribution, "Subject: Ikhana Fire Mission Status," NASA Dryden Flight Research Center, Edwards, CA, 19 July 2008.

[98] Email from Randy Herrin, Situation Unit Leader, PNW IMT 3, "Subject: Re: WSFM Cal Fire Mission 2 is on ground," Canyon Complex ICP, Chico, CA, July 19, 2008.

*Fires over the American River Fire Complex in Placer County, Calif., are seen as taken by the AMS wildfire scanner aboard the Ikhana,
July 8, 2008. The yellow areas depict active fire areas. The image provides officials not only with fire information but also with vegetation-
type and burned-area information.*
NASA/Google

over the Cascadel fire. Following a repeat set of data
passes, the pilots changed course to scan the fire from
a different angle.

Smoke filled the canyons in rugged, mountainous
terrain, visible in the optical cameras but transparent
to the AMS-Wildfire sensor. The science team recali-
brated the sensor as the pilots prepared to fly additional
data passes. "This is tight maneuvering space here,"
Pestana commented as he completed a turn.[99]

As the science team collected images and had

them automatically overlaid onto Google Earth imagery, Pestana flew the UAS just as he would any
other research aircraft. He maintained awareness of
aircraft systems and performance, communicated
with FAA controllers, maneuvered per instructions
from FAA controllers to facilitate the smooth flow of
other air traffic, and flew precision passes over the
fires. Finally, he turned toward Edwards Air Force
Base and – once inside restricted airspace – began a
descent for landing. The Ikhana touched down on the
runway 3.5 hours after takeoff, completing another
successful WSFM sortie.[100] In order to better acquaint

99 Ikhana flight plans (9/19/08) and notes taken during WSFM-
2008-03 by Peter W. Merlin, NASA Dryden Flight Research
Center, Edwards, CA, Sept. 19, 2008.

100 Ibid.

California Gov. Arnold Schwarzenegger, left, talks with NASA Ames Research Center director S. Pete Worden during the governor's visit to Ames on July 14, 2008. Schwarzenegger visited Ames for a behind-the-scenes tour and briefings about NASA's support to firefighters battling California wildfires.
NASA Ames Research Center/Eric James

state authorities with NASA's firefighting technology, Ames Research Center director Simon "Pete" Worden hosted California governor Schwarzenegger during a tour of Ames in which the AMS sensor and the Ikhana were the centerpieces. Representatives of the California Department of Forestry and Fire Protection (CAL FIRE) were also in attendance. Worden praised the cooperative firefighting effort as a great example of how the federal government and the state can work together in the face of natural and manmade disasters.

"The Ames Research Center here at Moffett Field and the Dryden Flight Research Center at Edwards Air Force Base, along with NASA's Goddard Space Flight Center in Greenbelt, Maryland, are each playing vital roles in this effort," said Worden. "We're also working with NASA's Jet Propulsion Laboratory

in Pasadena, the California Department of Forestry and Fire Protection, the Governor's Office of Emergency Services and the National Interagency Fire Center to help fight these numerous wildfires."[101]

Chief Del Walters, CAL FIRE executive officer, was equally ebullient. "I'm very excited about the technology and perhaps equally as excited about the partnership that's developing here," he said. "Having been a field firefighter for many years, I wish I had had this tool 20 years ago. You can only imagine the feeling of seeing a fire take off up a hill and lifting embers and the wind blowing. And you don't know, from where you're standing, whether it's gone over

[101] Press release, "Governor and NASA Highlight Infrared Scanning Technology Helping to Fight California's Wildfires," http://gov.ca.gov/speech/10186/, Office of the Governor, Sacramento, CA, July 14, 2008.

the next road, over the next hill, over the next creek and what's out in front of you. You have an idea, because you're used to the area that you're fighting fire in – although we fight fire all over the state, the Forest Service fights fire all over the nation, so you're not always in an area that's known to you. So to know if there's a community out there that's being threatened and that you need to stop what you're doing and change gears and employ different tactics is of tremendous value to the firefighters and to the community that we serve."[102]

---

[102] Ibid.

Chapter Four
AFTER THE FIRE

The Ikhana Western States Fire Mission flights of 2007 and 2008 resulted in a number of significant accomplishments and lessons. Multiple government agencies and corporate entities worked together to develop the capabilities of a UAS for use as a disaster response tool.

and Airborne Science programs and the Earth Science Technology Office developed the AMS-Wildfire sensor with the intent of demonstrating the capabilities during the WSFM and later transitioning those capabilities to operational agencies.

The WSFM project team repeatedly demonstrated

With smoke from the Lake Arrowhead area fires streaming in the background, NASA's Ikhana unmanned aircraft heads out on a wildfire-imaging mission.
NASA

WSFM accomplishments

The Western States UAS Fire Missions, carried out by team members from NASA Ames Research Center, NASA Dryden Flight Research Center, the U.S. Department of Agriculture Forest Service, the National Interagency Fire Center, NOAA, the FAA, and General Atomics Aeronautical Systems Inc., were a resounding success and a historic achievement in the field of unmanned aircraft technology.

In the first milestone of the project, NASA scientists developed improved imaging and communications processes for delivering near real-time information to firefighters. They realized that providing firefighters with critical information in a timely manner required an improvement in sensor technology as well as autonomous analysis and visualization tools that a NASA partnership could deliver. NASA's Applied Sciences

the utility and flexibility of using a UAS as an effective tool to aid disaster response personnel through the employment of various platform, sensor, and data dissemination technologies related to improving near real-time wildfire observations and intelligence gathering techniques. Each successive flight expanded capabilities of the previous missions for platform endurance and range, number of observations made, and flexibility in mission and sensing reconfiguration.

Team members worked with the FAA to safely and efficiently integrate the unmanned aircraft system into the national airspace. NASA pilots flew the Ikhana in close coordination with FAA air traffic controllers, allowing the aircraft to maintain safe separation from other aircraft.

WSFM project personnel developed extensive contingency management plans to minimize the risk to the aircraft and the public, including the negotiation of

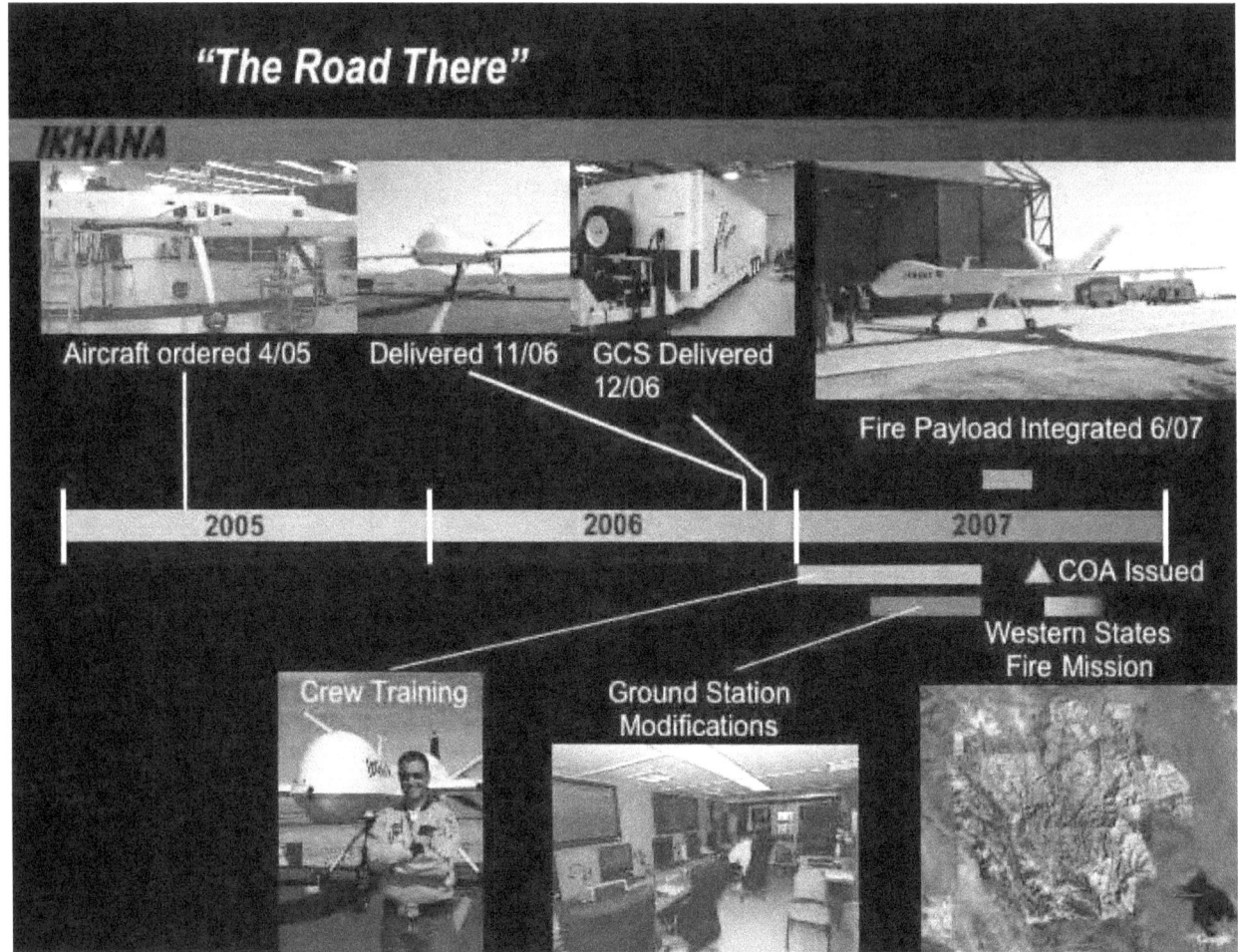

This timeline shows the major milestones of the Ikhana Western States Fire Mission project from 2005 through 2007.
NASA

emergency landing rights agreements at three government airfields and the identification and documentation of over 300 potential emergency landing sites in the event of an engine failure or other malfunction.

The missions included coverage of more than 60 wildfires throughout eight western states. All missions originated and terminated at Edwards Air Force Base and were operated by NASA crews with support from General Atomics. During the mission series, near real-time data was provided to Incident Command Teams and the National Interagency Fire Center.

Many fires were revisited during some missions to provide time-induced fire progression. Whenever possible, long-duration fire events were imaged on multiple missions to provide long-term fire monitoring capabilities. Post-fire burn-assessment imagery was also collected over various fires to aid teams in fire ecosystem rehabilitation. The project Flight Operations Team built relationships with other agencies

that enabled real-time flight plan changes necessary to avoid hazardous weather, to adapt to fire priorities, and to avoid conflicts with multiple planned military GPS testing/jamming activities.

Critical, near real-time fire information allowed Incident Command Teams to redeploy fire-fighting resources, assess effectiveness of containment operations, and move critical resources, personnel, and equipment from hazardous fire conditions. During instances where blinding smoke obscured normal observations, geo-rectified thermal-infrared data enabled the use of Geographic Information Systems or data visualization packages such as Google Earth. The images were collected and fully processed on board the Ikhana and transmitted via a communications satellite to NASA-Ames, where the imagery was served on a NASA Web site and provided in the Google Earth-based CDE for quick and easy access by incident commanders.

AMS imagery allowed incident commanders to chart the progress of wildfires and identify previously unknown hot spots.
NASA/Google

The Western States UAS Fire Mission series also gathered critical, coincident data with the EOS Terra and Aqua satellite sensor systems orbiting overhead, allowing for comparison and calibration of those resources with the more sensitive instruments on the Ikhana.

The WSFM team received an award for Group Achievement at the NASA Ames Research Center Honor Awards ceremony on Sept. 20, 2007. The team's award recognized excellence in providing operational near real-time fire condition information on the Esperanza fire from a UAS operating in the national airspace.[103]

Ikhana team members at Dryden and Ames were also recognized, along with the FAA and Forest Service, with a 2009 Federal Laboratory Consortium for Technology Transfer Interagency Partnership Award for collaborative use of unmanned aircraft to combat

forest fires. Dryden deputy director David D. McBride told team members, "The award you are receiving recognizes the contribution Ikhana has made to the transfer of UAS technology to civil applications." McBride also noted another benefit. "During this recession," he added, "it is important to appreciate the value technology has in moving the economy forward."[104]

The utility of near real-time data was dramatically illustrated on numerous occasions. During the July 2008 fire near Paradise, Calif., incident commanders were unaware of approaching danger because an undiscovered hotspot was concealed by smoke. After data from the Ikhana revealed the approaching flames, nearly 10,000 people were safely evacuated as firefighters allocated resources to protect the town. Lives and property were saved as a result. Firefighters were also able to use AMS-Wildfire data to study the dynamics and progression of various fires. In the short term, this allowed incident commanders to position firefighting assets in the most efficient manner. In the long term, it provided researchers with a valuable tool

[103] "Western States Fire Mission Team Award for Group Achievement," NASA Ames Research Center Honor Awards ceremony, NASA Ames Research Center, Mountain View, CA, Sept. 20, 2007, and Status Report, "NASA's Ikhana UAS Resumes Western States Fire Mission Flights," http://www.nasa.gov/centers/dryden/home/wsfm_status.html, NASA Dryden Flight Research Center, Edwards, CA, Sept. 19, 2008.

[104] Philip Hall, Brent Cobleigh, Greg Buoni, and Kathleen Howell, "Operational Experience with Long Duration Wildfire Mapping."

Post-burn imagery of the Zaca fire zone was collected as part of the Burned Area Emergency Rehabilitation project. Scientists used the data to examine the effects of fire on soil, watershed, wildlife, and vegetation. NASA/Google

for fire science studies.

The Ikhana UAS proved a versatile platform for carrying research payloads. Since the pod that housed the AMS-Wildfire sensor can be reconfigured, Ikhana could carry instruments for a variety of research projects. "There is interest in using Ikhana to study other types of phenomena such as weather, hurricanes, etc.," said Tom Rigney.[105]

Operational issues and lessons learned

Several significant operational issues were discovered during the WSFM project. The first of these was

[105] Interview with Thomas K. Rigney, Aug. 13, 2008.

a 2007 COA limitation to remain within 75 nmi of a backbone route. Because of this, the Ikhana science team could only investigate fires that were within the limited area. During the Aug. 30, 2007, flight major wildfires located in northern Idaho and Montana were beyond the approved operations area. FAA officials denied a NASA request to extend flights into this region, so only lower-priority fires were studied during that mission. The geographical restriction remained in effect through the end of the initial WSFM flight series because FAA officials wished to assess several Ikhana fire missions before extending the COA into new areas. Finally, during the 2008 Southern California Emergency Response Missions, the 75 nmi restriction was lifted and the Ikhana was permitted to fly beyond the boundaries initially specified in the 2007 COA. This gave mission planners greater flexibility and allowed more complete coverage of dynamic wildfire conditions.

An additional restriction to remain clear of regions of scheduled GPS testing initially did not appear significant even though planners had to schedule Ikhana flights around GPS testing/jamming exercises at military bases in the vicinity of WSFM routes. The FAA initiated this restriction because if the GCS were to lose the command link with the aircraft, the Ikhana's internal GPS and inertial navigation system would have to be used to fly a lost-link programmed route to return to Edwards airspace. During pre-mission planning for several flights, the NASA team had to coordinate directly with personnel at Nellis Air Force Base and China Lake Naval Warfare Center to keep Ikhana flight plans and GPS test operations at those sites from conflicting. In most cases, GPS testing had higher priority. On one occasion, an Ikhana flight was delayed 24 hours due to nearby GPS testing activities. During the California wildfire emergency, however, the WSFM team negotiated with Air Force authorities to give Ikhana flights priority over GPS testing/jamming exercises, thus allowing the mission to proceed. Affected regions were identified by FAA Notices to Airmen and typically consisted of an inverted cone centered at the test site and increasing in radius with increasing altitude. Although specifically how GPS testing/jamming might affect the Ikhana's navigation capabilities is unknown, it is thought that when flying at 25,000 feet, Ikhana performance could be affected at a range of up to 300 nmi from that test site.

Flight planning posed numerous challenges, beginning with the two-month long COA application process. Overall, the time spent preparing a complete and thorough application package paid off because, once submitted, there were few complications during the FAA review process. Face-to-face meetings and teleconferences between WSFM team members, FAA UAS Program Office personnel, and officials at affected Air Route Traffic Control Centers were invaluable in solving any problems with proposed flight plans and ensured a common understanding of the missions.

One weakness of the original COA application process was the fact that there was no single location at which the entire application resided after submission to the FAA. To complicate matters, when the FAA requested and received clarification from the WSFM team on specific matters, there was no guarantee that the initial application would be amended with the new information. This issue was resolved with the advent of a UAS COA online system to streamline the process by providing the applicant with a structured framework within which to answer questions and provide attachments with additional information.[106]

Because access to line-of-sight (LOS) communications frequencies was considered crucial, Ikhana mission planners had to carefully coordinate radio frequency management. For flight within approximately 70 nmi of Edwards, the Ikhana is controlled via a direct LOS radio link. The presence of significant military UAS operations in the vicinity required NASA planners to work around military flight schedules in order to have access to necessary radio frequencies. In many cases this meant that Ikhana flight operations in the local area were given lower priority and were therefore conducted outside of normal business hours. As a result, some of the Ikhana crew had to begin work as early as 3 a.m., further complicating scheduling issues for personnel that performed multiple duties on long-duration flights.

Unexpected weather along the flight route offered additional challenges. Each COA restricted flight from areas of adverse turbulence, convection, and icing. During the flight planning process it was difficult to account for weather variations, as these were inherently unpredictable. Flight tracks had to be designed and transmitted to FAA officials more than 72 hours in advance, meaning weather forecasts for the day of flight were not accurate to any meaningful extent. Weather forecasts, especially Airmen's Meteorological Information (AIRMETs) and Significant Meteorological Information (SIGMETs), were closely watched as the flight day approached but surprises sometimes occurred. The flight on Aug. 30, 2007, was launched with several convective SIGMETs issued for areas that were not along the planned route. Approximately two hours into the mission, the Ikhana was over Nevada when the weather service indicated the boundaries of a convective SIGMET along the flight track. The Ikhana team requested a significant heading deviation that took the

[106] Gregory P. Buoni and Kathleen M. Howell, "Large Unmanned Aircraft System Operations in the National Airspace System."

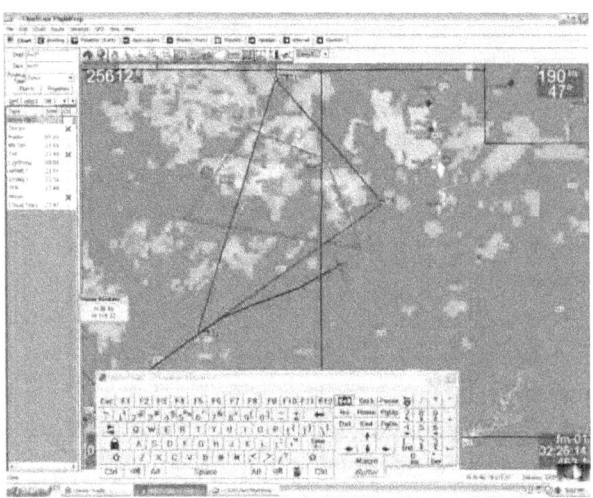

When a Significant Meteorological Information (SIGMET) bulletin indicated adverse weather along the Ikhana's flight route, air traffic controllers approved a heading deviation so the aircraft could avoid areas of high turbulence.
NASA/Google

aircraft several hundred miles from its original track to avoid the weather. Air traffic controllers approved the change, providing the necessary flexibility to allow the flight to continue safely.

FAA coordination for flights within the national airspace was absolutely crucial to WSFM objectives. Successful coordination of airspace access for Ikhana flights with FAA personnel was a groundbreaking achievement. Officials from both NASA and the FAA formed an effective partnership to plan and approve WSFM sorties. ARTCC personnel communicated their concerns and suggested resolutions whenever necessary. Thanks to the professionalism of both FAA and NASA staff members, no significant misunderstandings or miscommunications occurred. When the Ikhana was tasked to image fires during the emergency in Southern California, cooperation and coordination between the WSFM team and FAA was outstanding.

Staffing requirements, particularly for long-duration flights, created further challenges. Flights lasting longer than 10 hours required multiple crewmembers for all operational positions due to crew duty day limitations. This included pilots, system monitors, mission directors, and electronics and maintenance technicians. For flights longer than 12 hours, multiple shifts were implemented to comply with crew duty hour regulations. Since NASA did not have enough trained personnel to fill all duty stations, General Atomics was contracted to provide additional staff such as pilots and

Brent Cobleigh, left, the original Ikhana project manager, and his successor, Tom Rigney, oversaw NASA Dryden's participation in the Western States Fire Missions.
NASA

technicians. Non-standard flight schedules, intermittent sleep schedules, and extended on-call status have the potential to subject crewmembers to excessive fatigue. At crew briefings, project pilots made crew rest and readiness issues top priority in order to ensure that all members of the team were sufficiently rested and safety was not compromised.

Emergency response capabilities were tested during the Southern California firestorms. After conclusion of the last planned WSFM flight of 2007, the AMS-Wildfire pod was removed from the aircraft and preparations began for another experiment involving substantial modifications to the wing surfaces for sensor integration. When the California Office of Emergency Services made a request for Ikhana imagery, there were nearly a dozen active fires. More than 500,000 people had been evacuated, a number that would eventually double as the flames advanced. The Ikhana had to be reconfigured and the sensor pod reinstalled. The WSFM team quickly designed and submitted mission plans to the FAA with a request for an emergency COA. The keep-out zones were updated with re-evaluated population areas taking the evacuations into account. The emergency COA, based on the established WSFM COA, relieved some previous restrictions by allowing 24 rather than 72 hours' notice of a flight plan. The southern COA boundary was also extended to within 10 nmi of the Mexican border to allow greater sensor coverage. All of this demonstrated the ability of the Ikhana team to adapt quickly to a dynamic situation.[107]

Availability of contingency landing sites for the Ikhana was a concern that required a great deal of planning to provide successful risk management and to protect public safety. Primary landing sites included selected military airfields while secondary sites encompassed rural roads and open fields. Prior to the third 2008 WSFM sortie, a civilian at Air Combat Command (ACC) reviewed the emergency-landing plan for Mountain Home Air Force Base, Idaho, and convinced superiors to deny landing rights due to concerns about landing the Ikhana via satellite link. After a discussion with NASA officials allayed concerns, the base commander agreed to honor the original emergency landing rights agreement. When a new base commander was assigned to Mountain Home, however, he withdrew permission for landing based on the ACC recommendation. This action effectively eliminated parts of Northern California, Oregon, and Washington as operational areas until a different primary emergency landing site could be arranged. During the 2008 fire season, the NASA Ikhana team secured an agreement with the U.S. Army Yakima Training Center in Washington to designate an airfield for emergency use, thus enabling expansion of areas that could be covered during WSFM missions.[108]

Conclusions

The Western States Fire Mission was a tremendous success, enabling science teams and the fire management community to recognize that UAS platform and autonomous-operating sensor capabilities represent a viable solution to critical, real-time, disaster data gathering and support to national disaster management

communities. The Ikhana WSFM team demonstrated the effectiveness of using a UAS as a disaster response asset during multiple long-duration missions over active wildfires as well as for post-burn evaluation. Critical data was collected and disseminated to incident commanders in near real-time, allowing firefighters to more effectively employ their resources and initiate timely evacuation of threatened residents in affected areas. The WSFM team also demonstrated that a UAS could operate within the national airspace in the same way as a manned aircraft, accomplishing significant steps toward future "file and fly" missions.[109]

The Ikhana proved to be a versatile platform for the WSFM and other projects. The AMS scanner and CDE received strong praise from incident commanders at the U.S. Forest Service and National Interagency Fire Center, as well as from state and local fire teams. Most of all, hard work, dedication, and trust in each partner's key skills ensured the WSFM team's success. Teresa Fryberger, NASA's Director of Applied Science, called the team "a model for collaboration."[110]

Unmanned aircraft systems technology has the potential to bridge the gap between space-based and surface-based sensors and thus expand scientists' capabilities to monitor the global environment. Remotely piloted and autonomous platforms can provide critical coverage over remote and dangerous areas where manned aircraft flights are not practical due to long flight durations and hazardous conditions.

NASA remains at the forefront of this new realm and the Ikhana is just one of several unmanned vehicles in the Dryden fleet. Under a Space Act Agreement signed in May 2008, Dryden acquired two Northrop Grumman Global Hawk aircraft – the first civilian application of this autonomous, high-altitude, long-endurance UAS – for airborne science missions. The first is expected to fly for Dryden in 2009. And the center is also involved in a joint program with the Boeing Aircraft Company and Cranfield Aerospace, Ltd. of England to explore the characteristics and

[107] Philip Hall, Brent Cobleigh, Greg Buoni, and Kathleen Howell, "Operational Experience with Long Duration Wildfire Mapping."

[108] "Western States Fire Mission Tech Brief," NASA Dryden Flight Research Center, Sept. 26, 2008.

[109] Philip Hall, Brent Cobleigh, Greg Buoni, and Kathleen Howell, "Operational Experience with Long Duration Wildfire Mapping."

[110] "Western States Fire Mission Team Award for Group Achievement," NASA Ames Research Center Honor Awards ceremony, NASA Ames Research Center, Mountain View, CA, Sept. 20, 2007.

capabilities of a blended wing body aircraft. The X-48B is a scale, but wholly accurate, aircraft that is flown from a ground control station at Dryden. The test vehicle, powered by remote control modelers jet engines, is flown over Rogers Dry Lake but the pilot never actually sees the vehicle itself.

If unmanned aircraft systems are to evolve to support additional civil applications, pilot projects such as the Western States Fire Mission are critical to identifying and resolving technology shortcomings and developing operational procedures. The Ikhana and the WSFM paved the way for future civil UAS missions.

The Ikhana team was the project's most valuable asset. Hard work and dedication on the part of each member ensured mission success.
NASA

APPENDICES

Resources

This section includes a list of NASA Wildfire Response Team publications, articles, and technical papers, many of which provide additional information beyond the scope of this monograph.

Active fire mapping — airborne data

Ambrosia, V.G., S.S. Wegener, D.V. Sullivan, S.W. Buechel, J.A. Brass, S.E. Dunagan, R.G. Higgins, E.A. Hildum and S.M. Schoenung, 2002. Demonstrating UAV-Acquired Real-Time Thermal Data Over Fires, *Photogrammetric Engineering and Remote Sensing.*

Ambrosia, V.G., 2002. Emerging Technology Development For Disaster Management At NASA-Ames, Panel Session: Remote Sensing: New Technologies For Disaster Management, *Proceedings of the Twenty-Seventh Annual Hazards Research and Applications Workshop,* Boulder, Colorado, 14-17 July.

Ambrosia, V.G., 2002. Remotely Piloted Vehicles as Fire Imaging Platforms: The Future Is Here! *Wildfire Magazine,* May-June.

Wegener, S.S., V.G. Ambrosia, J. Stoneburner, D.V. Sullivan, J.A. Brass, S.W. Buechel, R.G. Higgins, E.A. Hildum, S.M. Schoenung, 2002. Demonstrating Acquisition of Real-Time Thermal Data Over Fires Utilizing UAVs. *Proceedings of AIAA's 1st Technical Conference and Workshop on Unmanned Aerospace Vehicles, Systems, Technologies, and Operations,* Portsmouth, Virginia, 20-23 May, Paper No. AIAA-2002-3406.

Ambrosia, V.G., S. S. Wegener, J.A. Brass and S.W. Buechel, 2002. Demonstrating Acquisition of Real-Time Thermal Data Over Fires Utilizing UAV's, RS2002, *Proceedings of the Ninth Biennial Remote Sensing Applications Conference,* San Diego, California, 8-12 April.

Brass, J.A., V.G. Ambrosia, R.S. Dann, R.G. Higgins, E.A. Hildum, J. McIntire, P.J. Riggan, S.M. Schoenung, R.E. Slye, D.V. Sullivan, S. Tolley, R. Vogler,

S.S. Wegener, 2001. First Response Experiment (FiRE) Using An Uninhabited Aerial Vehicle (UAV). *Proceedings of Fifth International Airborne Remote Sensing Conference,* San Francisco, California, 17-20 September, CD paper no. 56, pp 1-9.

Ambrosia, V.G., J.A. Brass, S.S. Wegener, D.V. Sullivan, S.W. Buechel and R.S. Dann, 2001. An Integration of Remote Sensing, Satellite Telemetry, and GIS Data Management Utilizing UAV Airborne Platforms For Application to Forest Fire Management. *Proceedings of the Third International Workshop: Remote Sensing and GIS Applications to Forest Fire Management: New Methods and Sensors,* European Association of Remote Sensing Laboratories (EARSeL), Paris, France, 17-19 May (abstract and poster).

Brass, J., V. Ambrosia, R. Higgins, T. Hildum, S. Schoenung, R. Slye, D. Sullivan, S. Tolley, H. Tran, R. Vogler, S. Wegener, 2000. Development of Tactical and Strategic Thermal Reconnaissance of Wildfires Utilizing Uninhabited Aerial Vehicle (UAV) Technology, *Proceedings, Fire Conference 2000: The First National Congress on Fire Ecology, Prevention and Management,* San Diego, CA., 27 Nov.-1 Dec., (abstract and poster).

Wegener, S., D. Sullivan, V. Ambrosia, J. Brass, R. Scott Dann, 2000. Development and Implementation of Real-Time Information Delivery Systems for Emergency Management. *Proceedings, First Int'l Global Disaster Information Network (GDIN) Information Technology Exposition and Conference,* Honolulu, HI, 9-11 Oct. 2000, (extended abstract).

Ambrosia, V., 1999. An Integration of Remote Sensing GIS, and Information Distribution For Wildfire Detection and Management. *Proceedings, The Joint Fire Science Conference and Workshop,* Vol. 1, p. 58-9, Boise, ID, 15-17 June, (abstract).

Ambrosia, V.G., S.W. Buechel, J.A. Brass, J.R. Peterson, R.H. Davies, R.J. Kane and S. Spain, 1998. An Integration of Remote Sensing, GIS, and Information Distribution for Wildfire Detection and Management. *Photogrammetric Engineering and Remote Sensing,* Vol. 64, No. 10, pp. 977-986.

Ambrosia, V.G., J.A. Brass, and R.G. Higgins, 1996. AIRDAS, Development of a Unique Four-Channel Scanner For Disaster Assessment and Management.

Proceedings of Second International Airborne Remote Sensing Conference and Exhibition, ERIM, Vol. III, 24-27 June, pp. 781-789.

Ambrosia, V.G., J.A. Brass, R.G. Higgins, and E.A. Hildum, 1996. Development and Utility of a Four-Channel Scanner For Wildfire Research and Applications. *Proceedings of the Sixth Biennial Forest Service Remote Sensing Applications Conference*, 29 April—3 May, pp. 390-399.

Ambrosia, V.G., J.A. Brass, J.B. Allen, E.A. Hildum, and R.G. Higgins, 1994. AIRDAS, Development of a Unique Four-Channel Scanner For Natural Disaster Assessment. *Proceedings of the First International Airborne Remote Sensing Conference and Exhibition*, ERIM, Vol. II, pp. 129-141.

Ambrosia, V.G., J.A. Brass, R.G. Higgins, P.J. Riggan, and R Lockwood, 1994. Development and Utility of a Four-Channel Scanner For Wildland Fire Research and Applications. *Proceedings of the Fifth Forest Service Remote Sensing Applications Conference*, pp. 311.

Ambrosia, V.G., 1990. High Altitude Aircraft Remote Sensing During the 1988 Yellowstone National Park Wildfires. *Geocarto International Journal*, V.5, No.3, pp. 43-47.

Ambrosia, V.G., 1990, High Altitude Forest Fire Mapping, *Daedalus Enterprises, Inc. Publication*. September 1990.

Ambrosia, V. G., J. A. Brass, R. Lathrop, K. Hibbard, P. Riggan, 1990. The Yellowstone Fire One Year Later -- Research in Remote Sensing Analysis. *Proceedings of the Association of American Geographers National Meeting*, Toronto, Canada, Program Abstracts, p. 3.

Airborne wildfire mapping

Ustin, S.L., Scheer, G., Castaneda, C.M., Jacquemoud, S., Roberts, D., and Green, R.O., 1996. Estimating Canopy Water Content of Chaparral Shrubs using Optical Methods, Summaries of the Sixth Annual JPL Airborne Earth Science Workshop, March 4-8, 1996. Vol. 1., AVIRIS Workshop, 4 pp

Ustin, S.L., Roberts, D.A., Pinzon, J., Jacquemoud, S., Gardner, M., Scheer, G., Castaneda, C.M. and Palacios, A., 1998, Estimating Canopy Water Content of Chaparral Shrubs Using Optical Methods, *Remote Sensing of Environment*. 65:280-291.

Roberts, D.A., Gardner, M., Regelbrugge, J., Pedreros, D. and Ustin, S., 1998. Mapping the distribution of wildfire fuels using AVIRIS in the Santa Monica Mountains, Proc. 7th AVIRIS Earth Science Workshop JPL 97-21, Pasadena, CA 91109, 345-352.

Serrano, L., Ustin, S.L., Roberts, D.A., Gamon, J.A., and Penuelas, J., 2000, Deriving water content of chaparral vegetation from AVIRIS data. *Remote Sensing of Environment*, 74: 570-581.

Dennison, P.E., Roberts, D.A., and Regelbrugge, J.C., 2000, Characterizing chaparral fuels using combined hyperspectral and synthetic aperture radar data, Proc. 9th AVIRIS Earth Science Workshop, JPL, Pasadena, CA Feb 23-25, 2000, 119-124

Brass, J.A. P.J. Riggan, V.G. Ambrosia, R.N. Lockwood, J.A. Pereira, R.G. Higgins, 1996, Brazil Fire Characterization and Burn Area Estimations Using the Airborne Infrared Disaster Assessment System. *Biomass Burning and Global Change*, ed. J. Levine, MIT Press, Vol. 2, Chapter 54, pp. 561-568.

Cofer, W.R., III, J.S. Levine, D.I. Sebacher, E.L. Winstead, P.J. Riggan, B.J. Stocks, J.A. Brass, V.G. Ambrosia and P.J. Boston, 1989, Trace Gas Emissions from Chaparral and Boreal Forest Fires, *Journal of Geophysical Research*, 94:2255.

Ambrosia, V.G. and J.A. Brass, 1989, Remote Sensing Thermal Analysis of the 1988 Yellowstone National Park Wildfires. *Association of American Geographers National Meeting*, Baltimore, Maryland, Program Abstracts, Sec. VI, p. 3.

Cofer, W.R., III, J.S. Levine, P.J. Riggan, D.I. Sebacher, E.L. Winstead, E.F. Shaw, Jr., J.A. Brass, and V.G. Ambrosia, 1988, Trace Gas Emissions from Mid-Latitude Prescribed Chaparral Fire, *Journal of Geophysical Research*, V.93, No. D2, pp. 1653-1658.

Ambrosia, V.G., and J.A. Brass, 1988, Thermal Analysis of Wildfires and Effects on Global Ecosystem Cycling. *Geocarto International Journal*, V.3, No. 1, pp. 29-39.

Cofer, W.R., III, J.S. Levine, P.J. Riggan, D.I. Sebacher, E.L. Winstead, P.J. Riggan, J.A. Brass, and V.G. Ambrosia, 1988, Particulate Emissions from a Mid-Latitude Prescribed Chaparral Fire, *Journal of Geophysical Research*, 93, No D2, 5207-5212.

Ambrosia, V.G. and J.A. Brass, 1988, Estimations of Thermal Gradients and Nutrient Cycling from Thermal Remote Sensing Data of Controlled Biomass Combustion. *Association of American Geographers National Meeting*, Phoenix, Arizona, Program Abstracts, p. 2, Sec. VII.

Brass, J.A., V.G. Ambrosia, and P.J. Riggan, 1988, Analysis of Fire/Fuel Characteristics as Related to Trace Gas Generation. *National Conference on Fire Management*, Los Angeles, California, April.

Ambrosia, V.G. and J.A. Brass, 1987, Thermal Analysis of Wildfires and Effect on Global Ecosystem Cycling. *Association of American Geographers National Meeting*, Portland, Oregon. Program Abstracts.

Brass, J.A., V.G. Ambrosia, and P.J. Riggan, 1987, Characterization of the Thermal Properties of Prescribed Wildland Fires Using Thermal Infrared Remote Sensing and Correlation to Nutrient Mobilization Due to Combustion. *Three Decades of Life Science Research in Space, Space Life Sciences Symposium*, Washington, DC, pp. 2, June.

Brass, J.A. and V.G. Ambrosia, 1987, Uses of Remote Sensing in Fire/Fuel Management. *Proceedings of the National Conference on Fire Management*, Los Angeles, California, June.

Brass, J.A., V.G. Ambrosia, P.J. Riggan, J.S. Myers, and J.C. Arvesen, 1987, Aircraft and Satellite Thermographic Systems for Wildfire Mapping and Assessment, *American Institute of Aeronautics and Astronautics*, AIAA-87-0187, January.

Brass, J.A. and V.G. Ambrosia, 1986, Uses of Remote Sensing in Fire/Fuel Management. *Proceedings of the National Conference on Fire Management*, Los Angeles, California, June.

Brass, J.A., V.G. Ambrosia, and P.J. Riggan, 1986. Biomass Combustion Characterization for Nutrient Movement Modeling Using Remotely Sensed Data. *Proceedings of the Ecological Society of America*, 67(2): 151, June 1986.

Brass, J.A., V.G. Ambrosia, and P. Riggan, 1985. Remotely Sensed Information of Biomass Combustion and Effects on Mobilization of Nutrients at the San Dimas Experimental Forest, *California. AIBS/ESA Conference, ESA Bulletin*, (ESA), 66(2): 146, pp. 1.

Satellite wildfire mapping

Brass, J.A., P.J. Riggan, V.G. Ambrosia, R.N. Lockwood, Raposo, J.A.P. and R.G. Higgins, 1997. Fires and Global Change: Prospects For Remote Sensing Assessment. In *World Resources Review*, Cambridge, MA, 10 pages.

Hlavka, C.A., V.G. Ambrosia, J.A. Brass, A. Rezendez, and L.S. Guild, 1996. Mapping Fire Scars in the Brazilian Cerrado Using AVHRR Imagery. *Biomass Burning and Global Change*, ed. J. Levine, MIT Press, Vol. 2, Chapter 53, pp. 550-560.

Guild, L.S., C.A. Hlavka, J.A. Brass, R.B. Chatfield, P.A. Matson, and V.G. Ambrosia, 1993, Biomass Burning In The Brazilian Cerrado: Early Results. *Bulletin of the Ecological Society of America*, 74(2), Addendum, 1993.

AMS algorithm development

Adams, J.B., Sabol, D., Kapos, V., Almeida Filho, R., Roberts, D.A., Smith, M.O., Gillespie, A.R., 1995, Classification of Multispectral Images Based on Fractions of Endmembers: Application to Land-Cover Change in the Brazilian Amazon, *Remote Sensing Environment*, 52:137-154.

Dennison, P., Roberts, D.A., Reith, E., Regelbrugge, J., and Ustin, S.L., 1999, Integrating Polarimetric Synthetic Aperture Radar and Imaging Spectrometry for Wildland fuel Mapping, Proc. of the Joint Fire Science Conference and Workshop, June 17-19, 1999, Boise, Idaho, Vol. 1, 37-42.

Roberts, D.A., Smith, M.O., Adams, J.B., Sabol, D.E., Gillespie, A.R., and Willis, S.C., 1990, Isolating Woody Plant Material and Senescent Vegetation from Green Vegetation in AVIRIS Data, *Proc. 2nd AVIRIS Workshop*, Pasadena, Ca., June 4 and 5, 1990, pp. 42-57.

Roberts, D.A., Smith, M.O., Sabol, D.E., Adams, J.B. and Ustin, S., 1992, Mapping the Spectral Variability in Photosynthetic and Non-Photosynthetic Vegetation, Soils and Shade using AVIRIS, *Summaries 3rd Annual JPL Airborne Geoscience Workshop*: Vol. 1, AVIRIS, Pasadena, CA. June 1 and 2, 1992, pp. 38-40.

Roberts, D.A., Green, R.O., Sabol, D.E. and Adams, J.B., 1993, Temporal Changes in Endmember Abundances, Liquid Water and Water Vapor over Vegetation at Jasper Ridge, *Summaries of the 4th Annual JPL Airborne Geoscience Workshop*, Oct 25-29, 1993, Vol. 1. AVIRIS Workshop, Washington D.C., 153-156.

Roberts, D.A., Adams, J.B., and Smith, M.O., 1993, Discriminating Green Vegetation, Non-Photosynthetic Vegetation and Soils in AVIRIS Data, *Rem. Sens. Environ.*, 44: 2/3 255-270.

Roberts, D.A., Green, R.O., Adams, J.B., Cothern, J.S., Sabol, D.E., and Smith, M.O., 1994, Temporal and Spatial Relationships Between Topography, Atmospheric Water Vapor, Liquid Water and Vegetation Endmember Fractions Determined Using AVIRIS. IGARRS '94, Pasadena, CA Aug 8-12, 1994, 3 p.

Roberts, D.A., Gardner, M., Church, R., Ustin, S., Scheer, G., and Green, R.O., 1996, Mapping Chaparral in the Santa Monica Mountains using Multiple Spectral Mixture Models, Summaries of the Sixth Annual JPL Airborne Earth Science Workshop, March 4-8, 1996. Vol. 1., AVIRIS Workshop, 5 pp

Roberts, D.A., Green, R.O., and Adams, J.B., 1997, Temporal and Spatial Patterns in Vegetation and Atmospheric Properties from AVIRIS, *Remote Sens. Environ.* 62: 223-240.

Roberts, D.A., Gardner, M., Church, R., Ustin, S.L., and Green, R.O., 1997, Optimum Strategies for Mapping Vegetation using Multiple Endmember Spectral Mixture Models, in SPIE Conf. Vol 3118, Imaging Spectrometry III, 108-119., San Diego, CA July 27-Aug 1, 1997.

Roberts, D.A., Brown, K.J., Green, R., Ustin, S., and Hinckley, T., 1998, Investigating the relationship between liquid water and leaf area in clonal Populus, Proc. 7th AVIRIS Earth Science Workshop JPL 97-21, Pasadena, CA 91109, 335-344.

Roberts, D.A., Gardner, M., Church, R., Ustin, S.,

Scheer, G., and Green, R.O., 1998, Mapping Chaparral in the Santa Monica Mountains using Multiple Endmember Spectral Mixture Models, *Rem. Sens. Environ.* 65: 267-279.

Roberts, D.A., Batista, G., Pereira, J., Waller, E., and Nelson, B. 1998, Change Identification using Multitemporal Spectral Mixture Analysis: Applications in Eastern Amazonia, Chapter 9 in Remote Sensing Change Detection: Environmental Monitoring Applications and Methods, (Elvidge, C. and Lunetta R., Eds.), Ann Arbor Press, Ann Arbor, MI, pp. 137-161.

Roberts, D.A., Dennison, P., Ustin, S.L., Reith, E., and Morais, M., 1999, Development of a Regionally Specific Library for the Santa Monica Mountains using High Resolution AVIRIS Data, Proc. 8th AVIRIS Earth Science Workshop, JPL, Pasadena, CA 349-354, Feb 8-11, 1999.

Roberts, D.A., Dennison, P.E., Morais, M., Gardner, M.E., Regelbrugge, J., and Ustin, S.L., 1999, Mapping Wildfire Fuels using Imaging Spectrometry along the Wildland Urban Interface, Proc. of the Joint Fire Science Conference and Workshop, June 17-19, 1999, Boise, Idaho, Vol. 1, 212-223.

Roberts, D.A., Numata, I., Holmes, K.W., Batista, G., Krug, T., Monteiro, A., Powell, B., and Chadwick, O., 2002, Large area mapping of land-cover change in Rondônia using multitemporal spectral mixture analysis and decision tree classifiers, *J. Geophys. Res. Atm.*, in press.

Rogan, J., Franklin, J., and Roberts, D.A., A comparison of methods for monitoring multitemporal vegetation change using Thematic Mapper imagery, 2002, *Remote Sensing of Environment*, 80 143-156.

Ustin, S.L., Scheer, G., Castaneda, C.M., Jacquemoud, S., Roberts, D., and Green, R.O., 1996, Estimating Canopy Water Content of Chaparral Shrubs using Optical Methods, Summaries of the Sixth Annual JPL Airborne Earth Science Workshop, March 4-8, 1996. Vol. 1., AVIRIS Workshop, 4 pp

Ustin, S.L., Roberts, D.A., Pinzon, J., Jacquemoud, S., Gardner, M., Scheer, G., Castaneda, C.M. and Palacios, A., 1998, Estimating Canopy Water Content of Chaparral Shrubs Using Optical Methods, *Rem. Sens.*

Environ. 65:280-291.

Integrated assessment of fire danger and fire spread modeling

Bossert, J.E. R.R. Linn, J.M. Reisner, J.L., Winterkamp, P. Dennison and D. Roberts, 2000, Coupled atmosphere-fire behavior model sensitivity to spatial fuels characterization, Third Symposium on Fire and Forest Meteorology, Amer. Meteor. Soc. 80th Annual Meeting, Long Beach, CA, 9-14 January 2000. pp. 21-26.

Fire effects, post-fire recovery, and hydrology

Ambrosia, V.G., J.A. Brass, P.J. Riggan, R. Ewing and P.D. Sebesta, 1997. Long-Term Stream Nitrate and Phosphate Changes Following Watershed Wildfires. *International Journal of Wildland Fire*, pp. 53-57.

Brass, J.A., V.G. Ambrosia, P.J. Riggan, and P.D. Sebesta, 1996. Consequences of Fire on Aquatic Nitrate and Phosphate Dynamics in Yellowstone National Park. *Ecological Implications of Fire in Greater Yellowstone*, IAWF, pp. 53-57.

Ambrosia, V.G., J.A. Brass, P.J. Riggan and P. Sebesta, 1994, Remote Sensing of Terrestrial and Aquatic Ecosystem Alterations Following the 1988 Yellowstone National Park Wildfires: Five Years of Research. *Proceedings of the Second Thematic Conference on Remote Sensing for Marine and Coastal Environments*, ERIM, Vol. II, pp. 243-256,

Loaiciga, H.A., Pedreros, D., and Roberts D.2001, Wildfire-stream flow interactions in a chaparral watershed, *Advances in Environmental Research*, 5 295-305.

Riano, D., Chuvieco, E., Ustin, S., Zomer, R., Dennison, P., Roberts, D., and Salas, J., 2002, Assessment of the vegetation regeneration after fire through the multitemporal analysis of AVIRIS images in the Santa Monica Mountains, *Remote Sensing of Environment*, 79(1), 60-71.

Sinha, R.P., J.A. Brass, V.G. Ambrosia and C.A. Hlavka, 1998, Global Warming and Burning of the Brazilian Cerrado. *Proceedings of the Fifth International Conference on Remote Sensing for Marine and*

Coastal Environments, ERIM, Vol. II, p. 505.

Souza, C. Jr., Firestone, L., Silva, L.M., and Roberts, D., Mapping forest degradation in the Eastern Amazon from SPOT 4 through spectral mixture models, *Remote Sensing of Environment.*

Education and outreach

Roberts, D.A., 1999, Camp on Fire, California Alliance for Minority Participation in Science, Engineering and Mathematics, Spring Quarterly, pp. 18-22.

Posada, Herman, 2008, NASA Western States Fire Mission, UVS International Conference, Paris France.

Hall, Philip, Brent Cobleigh, Greg Buoni, and Kathleen Howell, 2008, Operational Experience with Long Duration Wildfire Mapping UAS Missions over the Western United States, AUVSI-2008 Conference, San Diego, Calif.

Buoni, Gregory P. and Kathleen M. Howell, 2008, Large Unmanned Aircraft System Operations in the National Airspace System - the NASA 2007 Western States Fire

Missions, 8th AIAA Aviation Technology, Integration, and Operations Conference, 14-19 Sep. 2008, Anchorage, AK.

FAA Certification of UAS

Federal Aviation Administration

Memorandum

Date: November 15, 2006

From: Manager, Aircraft Engineering Division, AIR-100
       Manager, Production and Airworthiness Division, AIR-200

To: SEE DISTRIBUTION

Prepared by: James Sizemore, AIR-160, (202) 385-4631 or james.sizemore@faa.gov
             Richard Posey, AIR-230, (202) 267-9538 or richard.posey@faa.gov

Subject: Unmanned Aircraft Systems (UAS) Certification Status

This memorandum provides an update on the status of the efforts within AVS to allow Unmanned Aircraft Systems (UAS) to operate in the National Airspace System (NAS). In 2005, AIR-200 took the first step and began accepting applications for Special Airworthiness Certificates in the Experimental category for UAS. This activity is based on one of the strategies identified in the Administrator's Flight Plan for, "implementing technologies and systems that will help pilots operate aircraft as safely as possible." AIR-200 issued memorandums on June 17, 2005 and July 11, 2005 announcing this activity and AVS has subsequently issued 4 Experimental Certificates since August 2005.

The decision to manage UAS activity from FAA Headquarters is to gain experience with applicants that will allow us to develop policy and procedures for Aviation Safety Inspectors (ASI) and potentially for designees. AIR-200 has developed a draft order that describes the process for issuing Experimental Certificates to UAS and will prototype its use in FY 2007. All requests for an Experimental Certificate are processes by a team established within Headquarters composed of subject matter experts from AIR-100, AIR-200, AFS-400 and the Air Traffic Organization. Applicants are required to meet criteria established in a UAS unique Program Letter and Safety Checklist. AIR-200 works directly with the Manufacturing Inspection Offices, Manufacturing Inspection District/Satellite Office, or Certificate Management Offices/Units and Flight Standards District Offices to develop and finalize operating limitations, and issue Experimental Certificates. Program Letters or applications received by field offices should be referred to AIR-200.

The Unmanned Aircraft Program Office, AIR-160, was established in February 2006, and is the focal point for all AVS UAS activity, including any proposed certification projects that may be presented to ACO's. As such, all proposed UAS type certification projects must be worked within the appropriate Directorate, as well as coordinated with AIR-160.

There has been vigorous activity in industry exploring the endless possibilities that UAS bring to aviation. There are many new and novel issues to be addressed as UAS are considered for integration into the NAS. As these issues are unique to UAS and may differ substantially from manned aircraft operations and systems, they will require consideration and special attention while this area of aviation matures. At a minimum, the certification and introduction of these systems must not degrade the current level of safety within the NAS.

AIR-100 recently requested a point of contact (POC) from each Directorate to be the focal point to assist in the development of UAS specific policies, procedures, and potential regulations. These representatives will serve as the principle member for each directorate, to include associated field activities, in the development of necessary resources and milestones needed to accomplish programmatic tasks. These projects are significant, and early program coordination between the ACO Engineer and the Directorate POC is essential. As soon as practical after the project application, the ACO Engineer must notify and promptly forward the certification project notifications and associated certification plans to the Directorate POC. The Directorate POCs are responsible for ensuring coordination of these projects with AIR-160. In addition, as members of the Headquarters Team, it is anticipated that the Directorate POCs will participate in safety reviews with applicants for Experimental Certificates.

The list of the Directorate POC's are as follows:

Transport Airplane Directorate: Steve Edgar.
(425) 277 - 2025, steve.edgar@faa.gov

Small Airplane Directorate: Gunnar Berg and Greg Davison.
(816) 329 - 4141, gunnar.berg@faa.gov
(816) 329 - 4130, gregory.davison@faa.gov

Rotorcraft Directorate. Chinh Vuong.
(817) 222 - 5116, chinh.vuong@faa.gov

Engine and Propellor Directorate: Jeorge Fernandez.
(781) 238 - 7748, jeorge.fernandez@faa.gov

The AIR Headquarters POC's are as follows:

AIR-110 Victor Powell, Lead UAS Policy
(202) 267 - 9564, victor.powell@faa.gov

AIR-160 James Sizemore, Team Lead, UAS Airworthiness and Systems Engineering
(202) 385 - 4631, james.sizemore@faa.gov

AIR-200 Richard Posey, Lead UAS Experimental Certification Policy
(202) 267 - 9538, richard.posey@faa.gov

Thank you for your cooperation and support as we continue in this exciting new regulatory challenge in civil aviation. If there are any questions, please contact James Sizemore, AIR-160, and Richard Posey, AIR-230.

Distribution:
Director, Aircraft Certification Service
Director, Flight Standards Service
Manager, Brussels Aircraft Certification Staff, AEU-100
All Aircraft Certification Directorate Managers
All Aircraft Certification Offices
All Manufacturing Inspection Offices
All Manufacturing Inspection District/Satellite Offices
All Certificate Management Offices/Units
All Flight Standards Certificate Holding District Offices
Manager, Aircraft Maintenance Division, AFS-300
Manager, Flight Technologies and Procedures Division, AFS-400
Manager, General Aviation and Commercial Division, AFS-800

Ikhana WSFM preflight submission

This was the NASA Dryden Ikhana project L-1 day submission to the FAA for a COA authorizing WSFM-2, scheduled for July 19, 2008.

The following information is being provided to the FAA by the NASA Dryden Flight Research Center (DFRC) Ikhana project to fulfill Special Provision #4 of the FAA COA issued to NASA DFRC Operations and signed 7 July 2008. This notification will be followed by an IFR flight plan filing per the conditions of the COA.

The selected flight date is Saturday 7/19/2008 with take-off from KEDW at ~0800 PDT and planned landing at KEDW on Saturday 7/19/2008 at ~1500 PDT for a 7 hour flight.

The area of operations is wholly within R-2508/R-2515, and ZOA airspace with the route described below. All points in the NAS will be planned at FL230, unless otherwise directed by ATC per the conditions of the COA

Point/ Fire Name	FRD	Est Loiter Radius (nm)	Est Delay Time	Lat DMS	Lon DMS	Leg Dist	Total Dist
Within R-2508							
1	KEDW						0
2- 5 within R-2508 complex						37.7	95.1
6 FAANG	BIH189025			N 36.59.46.8	W118 34.45.6	98.2	193.3
In the NAS							
7 American River Complex	SWR247018	15	30 min	N 39.8.37.2	W120 38.43.8	161.7	354.9
8 Cub, Camp, Canyon Cplxs	CIC053023	30	90 min	N 39.55.40.8	W121 23.0	58.2	413.1
9 American River Complex	SWR247018	15	30 min	N 39.8.37.2	W120 38.43.8	58.2	471.3
10	MOD051026			N 37.47.10.2	W120 27.10.2	81.9	553.2
Within R-2508							
11 SWOOP	TTE025032			N 36.19.0	W118 35.4.8	125.7	678.9
12 – 15 within R-2508 complex							736.4
16	KEDW						831.4

Estimated time in ZOA airspace is from approximately 0930 – 1400 hours PDT

The fire locations, loiter radii and delay times are estimates and they may need to be adjusted somewhat during the flight to accommodate real-time fire conditions.

There appear to be less than 48 elements for an IFR flight plan in the NAS.

Per Special Provision 3 of the COA, the flight will be within 400 nm of identified Primary Emergency Landing Sites (ELS), and within 100 nm of secondary ELS.

Governor's statement

This statement was released to news media on July 14, 2008 from the Office of the California State Governor.

07/14/2008   GAAS:525:08   FOR IMMEDIATE RELEASE

**Governor Schwarzenegger and NASA Highlight Infrared Scanning Technology
Helping to Fight California's Wildfires**

Governor Arnold Schwarzenegger today joined NASA and federal and state fire officials at the NASA Ames Research Center in Moffett Field to tour the facility and discuss the important role of NASA's remotely piloted aircraft, named Ikhana, to California's firefight. The unmanned aircraft carrying a NASA infrared scanning sensor flew over much of California this past week, gathering information that was delivered to fire commanders in the field-helping them understand the terrain and behavior of the state's most dangerous fires. " California's unprecedented number of fires this early in the season make it all the more important that we use every tool at our disposal to protect property and save lives," Governor Schwarzenegger said. "NASA's Ikhana is one more incredible tool that we are able to use this year to bring real-time pictures and data to fire commanders, even when our other aircraft are unable to fly. The federal government has been an active partner in helping California fight fires, and NASA's assistance is one more example of that cooperation. " The Ikhana's most recent mission was on Tuesday, July 8. It flew for more than nine hours and covered approximately 10 individual and complex fires along a route over the Sierra Nevadas, west to the Cub Complex fires and south to the Gap Fire in Santa Barbara County. The images are collected onboard the Ikhana and transmitted through a communications satellite to NASA's Ames Research Center in Moffett Field, CA, where they are superimposed over Google Earth and Microsoft Virtual Earth maps to better visualize the location and scope of the fires. The imagery is then transmitted to the Multi-Agency Coordination Center in Redding and the State Operations Center in Sacramento, which distributes it to incident commanders in the field, so they can deploy resources where it will have the greatest benefit. NASA satellites are also capturing imagery of the wildfires to fill in gaps in airborne imagery. For these images and additional information, visit:

www.nasa.gov/topics/earth/features/fire_and_smoke.html.

BIBLIOGRAPHY

Adams, J.B., Sabol, D., Kapos, V., Almeida Filho, R., Roberts, D.A., Smith, M.O., Gillespie, A.R., "Classification of Multispectral Images Based on Fractions of Endmembers: Application to Land-Cover Change in the Brazilian Amazon." *Remote Sensing of Environment*, 1995.

"Altair Western States Fire Mission." http://ntrs.nasa.gov/archive/nasa/casi.ntrs.nasa.gov/20070031044_2007032019. pdf, accessed 10 June 2009.

"Altair / Predator B – An Earth Science Aircraft for the 21st Century." NASA Fact Sheet FS-073, NASA Dryden Flight Research Center, Edwards, CA, 2001.

"ALTAIR Unmanned Aircraft to Deploy to Canada." General Atomics Aeronautical Systems Company, San Diego, CA, 2004.

"ALTUS II – How High is High?" NASA Fact Sheet FS-1998-12-058 DFRC, NASA Dryden Flight Research Center, Edwards, CA, 1998.

Arjomandi, Maziar. "Classification of Unmanned Aerial Vehicles." Course material for Mechanical Engineering 3016, University of Adelaide, Australia, 2007.

Ambrosia, V.G. "High Altitude Forest Fire Mapping." *Daedalus Enterprises, Inc. Publication*, September 1990.

_____. "An Integration of Remote Sensing GIS, and Information Distribution For Wildfire Detection and Management." *Proceedings, The Joint Fire Science Conference and Workshop*, Vol. 1. Boise, ID, 15-17 June 1998 (abstract).

_____. "High Altitude Aircraft Remote Sensing During the 1988 Yellowstone National Park Wildfires." *Geocarto International Journal*, V.5, No.3, 1990.

_____. "Emerging Technology Development For Disaster Management At NASA-Ames, Panel Session: Remote Sensing: New Technologies For Disaster Management." *Proceedings of the Twenty-Seventh Annual Hazards Research and Applications Workshop, Boulder, Colorado*, 14-17 July 2002.

_____. "Remotely Piloted Vehicles as Fire Imaging Platforms: The Future Is Here!" *Wildfire Magazine*, May-June 2002.

Ambrosia, V. G., J. A. Brass, R. Lathrop, K. Hibbard, P. Riggan. "The Yellowstone Fire One Year Later -- Research in Remote Sensing Analysis." *Proceedings of the Association of American Geographers National Meeting*, Toronto, Canada, Program Abstracts, 1990.

Ambrosia, V.G. and J.A. Brass. "Remote Sensing Thermal Analysis of the 1988 Yellowstone National Park Wildfires." *Association of American Geographers National Meeting*, Baltimore, MD, Program Abstracts, Sec. VI, 1988.

_____. "Thermal Analysis of Wildfires and Effects on Global Ecosystem Cycling." *Geocarto International*

Journal, V.3, No. 1, 1988.

_____. "Estimations of Thermal Gradients and Nutrient Cycling from Thermal Remote Sensing Data of Controlled Biomass Combustion." *Association of American Geographers National Meeting*, Program Abstracts, Sec. VII Phoenix, AZ.

_____. "Thermal Analysis of Wildfires and Effect on Global Ecosystem Cycling." *Association of American Geographers National Meeting*, Program Abstracts, Portland, OR. 1987.

Ambrosia, V.G., J.A. Brass, P.J. Riggan, R. Ewing and P.D. Sebesta. "Long-Term Stream Nitrate and Phosphate Changes Following Watershed Wildfires." *International Journal of Wildland Fire*, 1997.

Ambrosia, V.G., J.A. Brass, P.J. Riggan and P. Sebesta. "Remote Sensing of Terrestrial and Aquatic Ecosystem Alterations Following the 1988 Yellowstone National Park Wildfires: Five Years of Research." *Proceedings of the Second Thematic Conference on Remote Sensing for Marine and Coastal Environments*, ERIM, Vol. II, 1994.

Ambrosia, V.G., J.A. Brass, S.S. Wegener, D.V. Sullivan, S.W. Buechel and R.S. Dann. "An Integration of Remote Sensing, Satellite Telemetry, and GIS Data Management Utilizing UAV Airborne Platforms For Application to Forest Fire Management." *Proceedings of the Third International Workshop: Remote Sensing and GIS Applications to Forest Fire Management: New Methods and Sensors*, European Association of Remote Sensing Laboratories (EARSeL), Paris, France, 17-19 May 2001.

Ambrosia, V.G., J.A. Brass, J.B. Allen, E.A. Hildum, and R.G. Higgins. "AIRDAS, Development of a Unique Four-Channel Scanner For Natural Disaster Assessment." *Proceedings of the First International Airborne Remote Sensing Conference and Exhibition*, ERIM, Vol. II, 1994.

Ambrosia, V.G., J.A. Brass, and R.G. Higgins. "AIRDAS, Development of a Unique Four-Channel Scanner For Disaster Assessment and Management." *Proceedings of Second International Airborne Remote Sensing Conference and Exhibition*, ERIM, Vol. III, 24-27 June 1996.

Ambrosia, V.G., J.A. Brass, R.G. Higgins, and E.A. Hildum. "Development and Utility of a Four-Channel Scanner For Wildfire Research and Applications." *Proceedings of the Sixth Biennial Forest Service Remote Sensing Applications Conference*, 29 April—3 May 1996.

Ambrosia, V.G., J.A. Brass, R.G. Higgins, P.J. Riggan, and R Lockwood." Development and Utility of a Four-Channel Scanner For Wildland Fire Research and Applications." *Proceedings of the Fifth Forest Service Remote Sensing Applications Conference*, 1994.

Ambrosia, V.G., S.W. Buechel, J.A. Brass, J.R. Peterson, R.H. Davies, R.J. Kane and S. Spain. "An Integration of Remote Sensing, GIS, and Information Distribution for Wildfire Detection and Management." *Photogrammetric Engineering and Remote Sensing*, Vol. 64, No. 10, 1998.

Ambrosia, V.G., S.S. Wegener, D.V. Sullivan, S.W. Buechel, J.A. Brass, S.E. Dunagan, R.G. Higgins, E.A. Hildum and S.M. Schoenung. "Demonstrating UAV-Acquired Real-Time Thermal Data Over Fires." *Photogrammetric Engineering and Remote Sensing*, 2002.

Ambrosia, V.G., S. S. Wegener, J.A. Brass and S.W. Buechel. "Demonstrating Acquisition of Real-Time Thermal

Data Over Fires Utilizing UAV's." RS2002, *Proceedings of the Ninth Biennial Remote Sensing Applications Conference*, San Diego, CA, 8-12 April, 2002.

"BAER: Burned Area Emergency Rehabilitation." Lytle Fire 2003, IncidentControl.com, http://www.incidentcontrol.com/lytlefire/b_a_e_r.htm, October 2003, accessed 10 June 2009.

Berlant, Daniel. "Unmanned aircraft is latest firefighting tool." Communique, California Department of Forestry and Fire Protection, Sacramento, CA, http://www.fire.ca.gov/communications/downloads/communique/2007_winter/unmanned.pdf. Winter 2007, accessed 10 June 2009.

Bolton , W. R. "Measurements of Radiation in the Atmosphere." *NASA Tech Briefs*, DRC-98-32, http://www.techbriefs.com/Briefs/Sep98/DRC9832.html, 1998, accessed 10 June 2009.

Bossert, J.E. R.R. Linn, J.M. Reisner, J.L., Winterkamp, P. Dennison and D. Roberts. "Coupled Atmosphere-Fire Behavior Model Sensitivity to Spatial Fuels Characterization." *Third Symposium on Fire and Forest Meteorology, American Meteorological Society 80th Annual Meeting*, Long Beach, CA, 9-14 January 2002.

Brass, J.A. and V.G. Ambrosia. "Uses of Remote Sensing in Fire/Fuel Management." *Proceedings of the National Conference on Fire Management*, Los Angeles, CA, June 1987.

Brass, J.A., V.G. Ambrosia, and P. Riggan. "Remotely Sensed Information of Biomass Combustion and Effects on Mobilization of Nutrients at the San Dimas Experimental Forest, California." *AIBS/ESA Conference, ESA Bulletin*, (ESA), 66(2): 1985.

_____. "Analysis of Fire/Fuel Characteristics as Related to Trace Gas Generation." *National Conference on Fire Management*, Los Angeles, CA, April 1988.

_____. "Biomass Combustion Characterization for Nutrient Movement Modeling Using Remotely Sensed Data." *Proceedings of the Ecological Society of America*, 67(2), June 1986.

_____. "Characterization of the Thermal Properties of Prescribed Wildland Fires Using Thermal Infrared Remote Sensing and Correlation to Nutrient Mobilization Due to Combustion." *Three Decades of Life Science Research in Space, Space Life Sciences Symposium*, Washington, DC, June 1987.

Brass, J.A., V.G. Ambrosia, P.J. Riggan, J.S. Myers, and J.C. Arvesen. "Aircraft and Satellite Thermographic Systems for Wildfire Mapping and Assessment." *American Institute of Aeronautics and Astronautics*, AIAA-87-0187, January1987.

Brass, J.A., V.G. Ambrosia, R.S. Dann, R.G. Higgins, E.A. Hildum, J. McIntire, P.J. Riggan, S.M. Schoenung, R.E. Slye, D.V. Sullivan, S. Tolley, R.Vogler, S.S. Wegener. "First Response Experiment (FiRE) Using An Uninhabited Aerial Vehicle (UAV)." *Proceedings of Fifth International Airborne Remote Sensing Conference*, San Francisco, CA, 17-20 September, CD paper no. 56, 2001.

Brass, J., V. Ambrosia, R. Higgins, T. Hildum, S. Schoenung, R. Slye, D. Sullivan, S. Tolley, H. Tran, R. Vogler, S. Wegener, 2000. Development of Tactical and Strategic Thermal "Reconnaissance of Wildfires Utilizing Uninhabited Aerial Vehicle (UAV) Technology." *Proceedings, Fire Conference 2000: The First National Congress on Fire Ecology, Prevention and Management*, San Diego, CA., 27 November-1 December, (abstract and poster).

Brass, J.A., V.G. Ambrosia, P.J. Riggan, and P.D. Sebesta. "Consequences of Fire on Aquatic Nitrate and Phosphate Dynamics in Yellowstone National Park." *Ecological Implications of Fire in Greater Yellowstone*, IAWF, 1996.

Brass, J.A., P.J. Riggan, V.G. Ambrosia, R.N. Lockwood, Raposo, J.A.P. and R.G. Higgins. "Fires and Global Change: Prospects For Remote Sensing Assessment." *World Resources Review*, Cambridge, MA, 1997.

Brass, J.A. P.J. Riggan, V.G. Ambrosia, R.N. Lockwood, J.A. Pereira, R.G. Higgins. "Brazil Fire Characterization and Burn Area Estimations Using the Airborne Infrared Disaster Assessment System." *Biomass Burning and Global Change*, J. Levine, ed., MIT Press, MA, Vol. 2, 1996.

Brown, Alan. "NASA's Newest Unmanned Aircraft Makes Successful First Flight." Press Release 03-193, NASA Headquarters, Washington, D.C., 2003.

Buoni, Gregory P. "Ikhana Weekly Notes." NASA Dryden Flight Research Center, Edwards, CA, 26 October 2008.

Buoni, Gregory P., and Kathleen M. Howell. "Large Unmanned Aircraft System Operations in the National Airspace System – the NASA 2007 Western States Fire Missions," AIAA-2008-8967, American Institute of Aeronautics and Astronautics, The 26th Congress of International Council of the Aeronautical Sciences, Anchorage, AK, 2008.

Chapman, William G. *Organizational Concepts for the Sensor-to-Shooter World – The Impact of Real-Time Information on Airpower Targeting*, Air University Press, Maxwell Air Force Base, AL, May 1997.

Cofer, W.R., III, J.S. Levine, P.J. Riggan, D.I. Sebacher, E.L. Winstead, E.F. Shaw, Jr., J.A. Brass, and V.G. Ambrosia. "Trace Gas Emissions from Mid-Latitude Prescribed Chaparral Fire." *Journal of Geophysical Research*, V.93, No. D2, 1988.

Cofer, W.R., III, J.S. Levine, D.I. Sebacher, E.L. Winstead, P.J. Riggan, B.J. Stocks, J.A. Brass, V.G. Ambrosia and P.J. Boston. "Trace Gas Emissions from Chaparral and Boreal Forest Fires." *Journal of Geophysical Research*, 1989.

"Completed Missions," Wildfire Research and Applications Partnership (WRAP), http://geo.arc.nasa.gov/sge/WRAP/current/com_missions.html, 2008, accessed 10 June 2009.

Dennison, P.E., Roberts, D.A., and Regelbrugge, J.C. "Characterizing Chaparral Fuels Using Combined Hyperspectral and Synthetic Aperture Radar Data." *Proceedings of the 9th AVIRIS Earth Science Workshop*, JPL, Pasadena, CA February 23-25, 2000.

Dennison, P., Roberts, D.A., Reith, E., Regelbrugge, J., and Ustin, S.L. "Integrating Polarimetric Synthetic Aperture Radar and Imaging Spectrometry for Wildland fuel Mapping." *Proceedings of the Joint Fire Science Conference and Workshop*, June 17-19, 1999, Boise, ID, Vol. 1, 1999.

"ERAST: Environmental Research and Sensor Technology Fact Sheet." NASA Dryden Flight Research Center, Edwards, CA, 2002.

Fahey, David W., James H. Churnside, James W. Elkins, Albin, J. Gasiewski, Karen H. Rosenlof, Sara Summers, Michael Aslaksen, Todd A. Jacobs, Jon D. Sellars, Christopher D. Jennison, Lawrence C. Freudinger, and Michael Cooper. "Altair Unmanned Aircraft System Achieves Demonstration Goals," *EOS Trans.*, No. 80. pp. 197-201,

American Geophysical Union, Washington, D.C., 16 May 2006.

"First ISCCP Regional Experiment (FIRE) Cirrus 2 NASA ER-2 Moderate Resolution Imaging Spectroradiometer (MODIS) Airborne Simulator (MAS) (FIRE_CI2_ER2_MAS) Langley DAAC Data Set Document" Atmospheric Science Data Center, NASA Langley Research Center, Hampton, VA, http://eosweb.larc.nasa.gov/GUIDE/dataset_documents/base_fire_ci2_er2_mas_dataset.html, 1996, accessed 10 June 2009.

Fulghum, David A., and Bill Sweetman. "Predator C Avenger Makes First Flights," *Aviation Week & Space Technology*, 17 April 2009.

Gaughan, Richard. "An autonomous sensor developed by NASA proves its worth in firefighting," *R&D Daily*, Issue 0701, January 2007.

"General Atomics Fact Sheet." General Atomics Aeronautical Systems Company, San Diego, CA, 2007.
"Governor and NASA Highlight Infrared Scanning Technology Helping to Fight California's Wildfires." http://gov.ca.gov/speech/10186/, Office of the Governor, Sacramento, CA, 14 July 2008.

"Ground Control Stations Fact Sheet." General Atomics Aeronautical Systems Company, San Diego, CA, 2007.

Guild, L.S., C.A. Hlavka, J.A. Brass, R.B. Chatfield, P.A. Matson, and V.G. Ambrosia. "Biomass Burning In The Brazilian Cerrado: Early Results." *Bulletin of the Ecological Society of America*, 74(2), Addendum, 1993.

Hagenauer, Beth. "NOAA and NASA Begin California UAV Flight Experiment." Press Release 05-20, NASA Dryden Flight Research Center, Edwards, CA, 2005.

_____. "Altair UAV Flies Lengthy Science Missions For NOAA," Photo Release 05-73P, NASA Dryden Flight Research Center, Edwards, CA, 2005.

_____. "Ikhana UAV Gives NASA New Science and Technology Capabilities." Press Release 07-12, NASA Dryden Flight Research Center, Edwards, CA, 2007.

Hagenauer, Beth, and Mike Mewhinney. "NASA Responds to California Wildfire Emergency Imaging Request," Release 08-30, NASA Dryden Flight Research Center, Edwards, CA, 11 July 2008.

Hall, Philip, Brent Cobleigh, Greg Buoni, and Kathleen Howell. "Operational Experience with Long Duration Wildfire Mapping UAS Missions over the Western United States," presented at the Association for Unmanned Vehicle Systems International Unmanned Systems North America Conference, San Diego, CA, June 2008.

Hlavka, C.A., V.G. Ambrosia, J.A. Brass, A. Rezendez, and L.S. Guild. "Mapping Fire Scars in the Brazilian Cerrado Using AVHRR Imagery." *Biomass Burning and Global Change*, J. Levine, ed., MIT Press, MA, Vol. 2, 1996.

"Ikhana Unmanned Science and Research Aircraft System." NASA Fact Sheet FS-097, NASA Dryden Flight Research Center, Edwards, CA, 2007.
Levine, Jay. "Measuring up to the Gold Standard." *X-tra*, NASA Dryden Flight Research Center, Edwards, CA, 2008.

_____. "No one on board – Ikhana pilots fly aircraft from the ground." *X-tra*, NASA Dryden Flight Research

Center, Edwards, CA, 2008.

Loaiciga, H.A., Pedreros, D., and Roberts D. "Wildfire-stream Flow Interactions in a Chaparral Watershed." *Advances in Environmental Research*, 2001.

McDaid, Hugh, and David Oliver. *Smart Weapons: Top Secret History of Remote Controlled Airborne Weapons*, Orion Media, London, England, 1997.

Miller, Jay. Lockheed U-2, Aerofax Inc., Austin, TX, 1983. Missions, 8th American Institute for Astronautics and Aeronautics Aviation Technology, Integration, and Operations Conference, 14-19 September 2008, Anchorage, AK.

"NASA Flies Ikhana UAV to Help California Firefighter." Aviation.com, http://www.aviation.com/technology/071024-nasa-ikhana-california-wildfires.html October 2007, accessed 10 June 2009.

Office of the Secretary of Defense. "Unmanned Aircraft Systems Roadmap 2005-2030," http://www.acq.osd.mil/usd/Roadmap%20Final2.pdf, 2005, accessed 12 October 2008.

Posada, Herman. "NASA Western States Fire Mission, UVS International Conference." Docks Event Centre, Paris, France, 2008.

Riano, D., Chuvieco, E., Ustin, S., Zomer, R., Dennison, P., Roberts, D., and Salas, J. "Assessment of the Vegetation Regeneration After Fire Through the Multitemporal Analysis of AVIRIS Images in the Santa Monica Mountains." *Remote Sensing of Environment*, 79(1), 2002.

Roberts, D.A. "Camp on Fire, California Alliance for Minority Participation in Science." *Engineering and Mathematics*, Spring Quarterly, 1999.

Roberts, D.A., Adams, J.B., and Smith, M.O. "Discriminating Green Vegetation, Non-Photosynthetic Vegetation and Soils in AVIRIS Data." *Remote Sensing Environment*, 44: 2/3, 1993.

Roberts, D.A., Dennison, P., Ustin, S.L., Reith, E., and Morais, M. "Development of a Regionally Specific Library for the Santa Monica Mountains using High Resolution AVIRIS Data." P*roceedings of the 8th AVIRIS Earth Science Workshop*, JPL, Pasadena, CA, February 8-11, 1999.

Roberts, D.A., Dennison, P.E., Morais, M., Gardner, M.E., Regelbrugge, J., and Ustin, S.L. "Mapping Wildfire Fuels using Imaging Spectrometry along the Wildland Urban Interface." *Proceedings of the Joint Fire Science Conference and Workshop*, June 17-19, 1999, Boise, ID, Vol. 1.

Roberts, D.A., Brown, K.J., Green, R., Ustin, S., and Hinckley, T. "Investigating the Relationship Between Liquid Water and Leaf Area in Clonal Populus." *Proceedings of the 7th AVIRIS Earth Science Workshop*, JPL, Pasadena, CA, 1998.

Roberts, D.A., Batista, G., Pereira, J., Waller, E., and Nelson, B. "Change Identification using Multitemporal Spectral Mixture Analysis: Applications in Eastern Amazonia." Chapter 9 in *Remote Sensing Change Detection: Environmental Monitoring Applications and Methods*, (Elvidge, C. and Lunetta R., Eds.), Ann Arbor Press, Ann Arbor, MI, 1998.

Roberts, D.A., Gardner, M., Church, R., Ustin, S., Scheer, G., and Green, R.O. "Mapping Chaparral in the Santa Monica Mountains using Multiple Spectral Mixture Models." *Summaries of the Sixth Annual JPL Airborne Earth Science Workshop*, Vol. 1, March 4-8, 1996.

Roberts, D.A., Gardner, M., Church, R., Ustin, S.L., and Green, R.O. "Optimum Strategies for Mapping Vegetation using Multiple Endmember Spectral Mixture Models." SPIE Conference. Vol. 3118, *Imaging Spectrometry*, San Diego, CA July 27-Aug 1, 1997.

Roberts, D.A., Gardner, M., Church, R., Ustin, S., Scheer, G., and Green, R.O. "Mapping Chaparral in the Santa Monica Mountains using Multiple Endmember Spectral Mixture Models." *Remote Sensing of Environment*, 65, 1998.

Roberts, D.A., Gardner, M., Regelbrugge, J., Pedreros, D. and Ustin, S. "Mapping the distribution of wildfire fuels using AVIRIS in the Santa Monica Mountains." *Proceedings of the 7th AVIRIS Earth Science Workshop* JPL, Pasadena, CA, 1998.

Roberts, D.A., Green, R.O., and Adams, J.B. "Temporal and Spatial Patterns in Vegetation and Atmospheric Properties from AVIRIS." *Remote Sensing of Environment*, 62, 1997.

Roberts, D.A., Green, R.O., Adams, J.B., Cothern, J.S., Sabol, D.E., and Smith, M.O. "Temporal and Spatial Relationships Between Topography, Atmospheric Water Vapor, Liquid Water and Vegetation Endmember Fractions Determined Using AVIRIS." IGARRS '94, Pasadena, CA August 8-12, 1994.

Roberts, D.A., Green, R.O., Sabol, D.E. and Adams, J.B. "Temporal Changes in Endmember Abundances, Liquid Water and Water Vapor over Vegetation at Jasper Ridge." *Summaries of the 4th Annual JPL Airborne Geoscience AVIRIS Workshop*, October 25-29, 1993, Vol. 1 , Washington D.C.

Roberts, D.A., Numata, I., Holmes, K.W., Batista, G., Krug, T., Monteiro, A., Powell, B., and Chadwick, O. "Large Area Mapping of Land-Cover Change in Rondônia Using Multitemporal Spectral Mixture Analysis and Decision Tree Classifiers." J. *Geophys. Res. Atm.*, 2002, in press.

Roberts, D.A., Smith, M.O., Adams, J.B., Sabol, D.E., Gillespie, A.R., and Willis, S.C. "Isolating Woody Plant Material and Senescent Vegetation from Green Vegetation in AVIRIS Data." *Proceedings of the 2nd AVIRIS Workshop*, Pasadena, Ca., June 4-5, 1990.

Roberts, D.A., Smith, M.O., Sabol, D.E., Adams, J.B. and Ustin, S. "Mapping the Spectral Variability in Photosynthetic and Non Photosynthetic Vegetation, Soils and Shade using AVIRIS." *Summaries 3rd Annual JPL Airborne Geoscience Workshop*, Vol. 1, Pasadena, CA. June 1 and 2, 1992.

"Rocket launch secures photos for firefighting efforts." Alaska Science Outreach, http://www.alaskascienceoutreach. com/index.php/main_pages/catchitem/rocket_launch_secures_photos_for_firefighting_efforts/, accessed 10 June 2004.
Rogan, J., Franklin, J., and Roberts, D.A. "A Comparison of Methods for Monitoring Multitemporal Vegetation Change Using Thematic Mapper Imagery." *Remote Sensing of Environment*, 80, 2002.

Serrano, L., Ustin, S.L., Roberts, D.A., Gamon, J.A., and Penuelas, J. "Deriving Water Content of Chaparral Vegetation from AVIRIS Data." *Remote Sensing of Environment*, 74, 2000.

Sinha, R.P., J.A. Brass, V.G. Ambrosia and C.A. Hlavka. "Global Warming and Burning of the Brazilian Cerrado." *Proceedings of the Fifth International Conference on Remote Sensing for Marine and Coastal Environments,* ERIM, Vol. II, 1998.

Souza, C. Jr., Firestone, L., Silva, L.M., and Roberts, D. "Mapping forest degradation in the Eastern Amazon from SPOT 4 through spectral mixture models." *Remote Sensing of Environment,* 2000.

Status Report, "NASA's Ikhana UAS Resumes Western States Fire Mission Flights," http://www.nasa.gov/centers/dryden/home/wsfm_status.html, NASA Dryden Flight Research Center, Edwards, CA, 19 September 2008.

Ustin, S.L., Roberts, D.A., Pinzon, J., Jacquemoud, S., Gardner, M., Scheer, G., Castaneda, C.M. and Palacios, A. "Estimating Canopy Water Content of Chaparral Shrubs Using Optical Methods." *Remote Sensing of Environment,* 65, 1998.

Ustin, S.L., Scheer, G., Castaneda, C.M., Jacquemoud, S., Roberts, D., and Green, R.O. "Estimating Canopy Water Content of Chaparral Shrubs using Optical Methods." *Summaries of the Sixth Annual JPL Airborne Earth Science AVIRIS Workshop,* March 4-8, 1996. Vol. 1.

Wegener, S.S., V.G. Ambrosia, J. Stoneburner, D.V. Sullivan, J.A. Brass, S.W. Buechel, R.G. Higgins, E.A. Hildum, S.M. Schoenung. "Demonstrating Acquisition of Real-Time Thermal Data Over Fires Utilizing UAVs." *Proceedings of AIAA's 1st Technical Conference and Workshop on Unmanned Aerospace Vehicles, Systems, Technologies, and Operations,* Portsmouth, Virginia, 20-23 May, Paper No. AIAA-2002-3406, 2002.

Wegener, S., D. Sullivan, V. Ambrosia, J. Brass, R. Scott Dann. "Development and Implementation of Real-Time Information Delivery Systems for Emergency Management." *Proceedings, First International Global Disaster Information Network (GDIN) Information Technology Exposition and Conference,* Honolulu, HI, 9-11 October 2000, (extended abstract).

"Western States Fire Mission Team Award for Group Achievement," NASA Ames Research Center Honor Awards ceremony, NASA Ames Research Center, Mountain View, CA, Sept. 20, 2007.

Wilhite, Jamie, Robert Navarro, and Brent Cobleigh. "Altair Western States Fire Mission." *2006 Engineering Annual Report,* NASA Dryden Flight Research Center, Edwards, CA, August 2007.

Unpublished Sources

Interviews with the author:
Vincent G. Ambrosia. 13 August 2008.
Mark Pestana 13 August 2008
Henana Posada 19 September 2008.
Thomas K. Rigney, 13 August 2008
Cobleigh, Brent. Ikhana Flight Reports. NASA Dryden Flight Research Center, Edwards, CA, 2007.

Herrin, Randy. Situation Unit Leader, PNW IMT 3, email, "Subject: Re: WSFM Cal Fire Mission 2 is on ground." Canyon Complex ICP, Chico, CA, 19 July 2008.

Ikhana flight plans (6/29/08) and meeting notes, Ikhana Team Meeting. NASA Dryden Flight Research Center,

Edwards, CA, 3 July 2008.

Ikhana flight plans (9/19/08) and notes taken during WSFM-2008-03 by Peter W. Merlin, NASA Dryden Flight Research Center, Edwards, CA, 19 September 2008.

Rigney, Thomas K. Ikhana Project Manager, email, "Subject: Ikhana Fire Mission Status." NASA Dryden Flight Research Center, Edwards, CA, 19 July 2008.

"Western States Fire Mission Tech Brief," NASA Dryden Flight Research Center, 26 September 2008.

INDEX

Documentary Histories

Exploring the Unknown

Logsdon, John M., ed., with Linda J. Lear, Jannelle Warren Findley, Ray A. Williamson, and Dwayne A. Day. *Exploring the Unknown: Selected Documents in the History of the U.S. Civil Space Program, Volume I, Organizing for Exploration.* NASA SP-4407, 1995.

Logsdon, John M., ed, with Dwayne A. Day, and Roger D. Launius. *Exploring the Unknown: Selected Documents in the History of the U.S. Civil Space Program, Volume II, External Relationships.* NASA SP-4407, 1996.

Logsdon, John M., ed., with Roger D. Launius, David H. Onkst, and Stephen J. Garber. *Exploring the Unknown: Selected Documents in the History of the U.S. Civil Space Program, Volume III, Using Space.* NASA SP-4407, 1998.

Logsdon, John M., ed., with Ray A. Williamson, Roger D. Launius, Russell J. Acker, Stephen J. Garber, and Jonathan L. Friedman. *Exploring the Unknown: Selected Documents in the History of the U.S. Civil Space Program, Volume IV, Accessing Space.* NASA SP-4407, 1999.

Logsdon, John M., ed., with Amy Paige Snyder, Roger D. Launius, Stephen J. Garber, and Regan Anne Newport. *Exploring the Unknown: Selected Documents in the History of the U.S. Civil Space Program, Volume V, Exploring the Cosmos.* NASA SP-4407, 2001.

Logsdon, John M., ed., with Stephen J. Garber, Roger D. Launius, and Ray A. Williamson. *Exploring the Unknown: Selected Documents in the History of the U.S. Civil Space Program, Volume VI: Space and Earth Science.* NASA SP-2004-4407, 2004.

The Wind and Beyond

Hansen, James R., ed. *The Wind and Beyond: Journey into the History of Aerodynamics in America, Volume 1, The Ascent of the Airplane.* NASA SP-2003-4409, 2003.

Hansen, James R., ed. *The Wind and Beyond: Journey into the History of Aerodynamics in America, Volume 2, Reinventing the Airplane.* NASA SP-2007-4409, 2007.

Brief Histories of NASA

Anderson, Frank W., Jr. *Orders of Magnitude: A History of NACA and NASA, 1915-1980.* NASA SP-4403, 1981.

Bilstein, Roger E. *Orders of Magnitude: A History of the NACA and NASA, 1915-1990.* NASA SP-4406, 1989.

Bilstein, Roger E. *Testing Aircraft, Exploring Space: An Illustrated History of NACA and NASA.* Baltimore: Johns Hopkins University Press, 2003.

Critical Issues in the History of Spaceflight

Dick, Steven J. and Launius, Roger D. *Critical Issues in the History of Spaceflight.* (NASA SP-2006-4702).

Societal Impact of Spaceflight

Dick, Steven J. and Launius, Roger D. *Societal Impact of Spaceflight.* (NASA SP-2007-4801).

Memoirs

Chertok, Boris. *Rockets and People,* Volume 1. (NASA SP-2005-4110). Visit http://history.nasa.gov/series95.html for a pdf version of this document.

Chertok, Boris. *Rockets and People: Creating a Rocket Industry,* Volume II. (NASA SP-2006-4110).

Mudgway, Douglas J. William H. Pickering: America's Deep Space Pioneer. (NASA SP-2007-4113).

Aeronautics and Space Report of the President

The annual "President's Report" is a summary of the Government's aerospace activities each year. Mandated by law, it contains information on aerospace activities conducted by 14 Federal departments and agencies. It also contains an executive summary organized by agency, narrative sections organized by subject, as well as extensive appendices containing useful historical data on spacecraft launches, budget figures, key policy documents from the fiscal year, and a glossary. Visit http://history.nasa.gov/series95.html for pdf versions of these documents.

NASA Historical Data Books

Van Nimmen, Jane, and Leonard C. Bruno, with Robert L. Rosholt. *NASA Historical Data Book, Vol. I: NASA Resources, 1958-1968.* NASA SP-4012, 1976, rep. ed. 1988.

Ezell, Linda Neuman. *NASA Historical Data Book, Vol. II: Programs and Projects, 1958-1968.* NASA SP-4012, 1988.

Ezell, Linda Neuman. *NASA Historical Data Book, Vol. III: Programs and Projects, 1969-1978.* NASA SP-4012, 1988.

Gawdiak, Ihor, with Helen Fedor. *NASA Historical Data Book, Vol. IV: NASA Resources, 1969 1978.* NASA SP-4012, 1994.

Rumerman, Judy A. *NASA Historical Data Book, Vol. V: NASA Launch Systems, Space Transportation, Human Spaceflight, and Space Science, 1979-1988.* NASA SP-4012, 1999.

Rumerman, Judy A. *NASA Historical Data Book, Vol. VI: NASA Space Applications, Aeronautics and Space Research and Technology, Tracking and Data Acquisition/Support Operations, Commercial Programs, and Resources, 1979-1988.* NASA SP-4012, 1999.

Astronautics and Aeronautics Chronology

Eugene M. Emme, comp. *Aeronautics and Astronautics Chronology, 1915-1960. Aeronautics and Astronautics: An American Chronology of Science and Technology in the Exploration of Space, 1915-1960* (Washington, DC:

National Aeronautics and Space Administration, 1961). Visit http://history.nasa.gov/series95.html for a pdf version of this document.

Eugene M. Emme, comp. *Aeronautical and Astronautical Events of 1961. Report of the National Aeronautics and Space Administration to the Committee on Science and Astronautics, U.S. House of Representatives, 87th Cong., 2d. Sess.* (Washington, DC: U.S. Government Printing Office, 1962). Visit http://history.nasa.gov/series95.html for a pdf version of this document.

Astronautical and Aeronautical Events of 1962. Report to the Committee on Science and Astronautics, Report to the Committee on Science and Astronautics, U.S. House of Representatives, Eighty-eighth Congress, first session (Washington, DC: U.S. Government Printing Office, 1963).

Astronautics and Aeronautics, 1963: Chronology of Science, Technology, and Policy. NASA SP-4004, 1964.

Astronautics and Aeronautics, 1964: Chronology of Science, Technology, and Policy. NASA SP-4005, 1965.

Astronautics and Aeronautics, 1965: Chronology of Science, Technology, and Policy. NASA SP-4006, 1966.

Astronautics and Aeronautics, 1966: Chronology of Science, Technology, and Policy. NASA SP-4007, 1967.

Astronautics and Aeronautics, 1967: Chronology of Science, Technology, and Policy. NASA SP-4008, 1968.

Astronautics and Aeronautics, 1968: Chronology of Science, Technology, and Policy. NASA SP-4010, 1969.

Astronautics and Aeronautics, 1969: Chronology of Science, Technology, and Policy. NASA SP-4014, 1970.

Astronautics and Aeronautics, 1970: Chronology of Science, Technology, and Policy. NASA SP-4015, 1972.

Astronautics and Aeronautics, 1971: Chronology of Science, Technology, and Policy. NASA SP-4016, 1972.

Astronautics and Aeronautics, 1972: Chronology of Science, Technology, and Policy. NASA SP-4017, 1974.

Astronautics and Aeronautics, 1973: Chronology of Science, Technology, and Policy. NASA SP-4018, 1975.

Astronautics and Aeronautics, 1974: Chronology of Science, Technology, and Policy. NASA SP-4019, 1977.

Astronautics and Aeronautics, 1975: Chronology of Science, Technology, and Policy. NASA SP-4020, 1979.

Astronautics and Aeronautics, 1976: Chronology of Science, Technology, and Policy. NASA SP-4021, 1984.

Astronautics and Aeronautics, 1977: Chronology of Science, Technology, and Policy. NASA SP-4022, 1986.

Astronautics and Aeronautics, 1978: Chronology of Science, Technology, and Policy. NASA SP-4023, 1986.

Astronautics and Aeronautics, 1979-1984: Chronology of Science, Technology, and Policy. NASA SP-4024, 1988.

Astronautics and Aeronautics, 1985: Chronology of Science, Technology, and Policy. NASA SP-4025, 1990.

Gawdiak, Ihor Y., Ramon J. Miro, and Sam Stueland, comps. *Astronautics and Aeronautics, 1986-1990: A Chronology.* NASA SP-4027, 1997.

Gawdiak, Ihor Y. and Shetland, Charles. *Astronautics and Aeronautics, 1991-1995: A Chronology.* NASA SP-2000-4028, 2000.

NASA Publications by Special Publication (SP) Numbers

Reference Works, NASA SP-4000

Grimwood, James M. *Project Mercury: A Chronology.* NASA SP-4001, 1963.

Grimwood, James M., and Barton C. Hacker, with Peter J. Vorzimmer. *Project Gemini Technology and Operations: A Chronology.* NASA SP-4002, 1969.

Link, Mae Mills. *Space Medicine in Project Mercury.* NASA SP-4003, 1965.

Ertel, Ivan D., and Mary Louise Morse. *The Apollo Spacecraft: A Chronology, Volume I, Through November 7, 1962.* NASA SP-4009, 1969.

Morse, Mary Louise, and Jean Kernahan Bays. *The Apollo Spacecraft: A Chronology, Volume II, November 8, 1962-September 30, 1964.* NASA SP-4009, 1973.

Brooks, Courtney G., and Ivan D. Ertel. *The Apollo Spacecraft: A Chronology, Volume III, October 1, 1964-January 20, 1966.* NASA SP-4009, 1973.

Ertel, Ivan D., and Roland W. Newkirk, with Courtney G. Brooks. *The Apollo Spacecraft: A Chronology,* Volume IV, January 21, 1966-July 13, 1974. NASA SP-4009, 1978.

Newkirk, Roland W., and Ivan D. Ertel, with Courtney G. Brooks. *Skylab: A Chronology.* NASA SP-4011, 1977.

Van Nimmen, Jane, and Leonard C. Bruno, with Robert L. Rosholt. *NASA Historical Data Book,* Vol. *I: NASA Resources, 1958-1968.* NASA SP-4012, 1976, rep. ed. 1988.

Ezell, Linda Neuman. *NASA Historical Data Book, Vol. II: Programs and Projects, 1958-1968.* NASA SP-4012, 1988.

Ezell, Linda Neuman. *NASA Historical Data Book, Vol. III: Programs and Projects, 1969-1978.* NASA SP-4012, 1988.

Gawdiak, Ihor, with Helen Fedor. *NASA Historical Data Book, Vol. IV: NASA Resources, 1969-1978.* NASA SP-4012, 1994.

Rumerman, Judy A. *NASA Historical Data Book, Vol. V: NASA Launch Systems, Space Transportation, Human Spaceflight, and Space Science, 1979-1988.* NASA SP-4012, 1999.

Rumerman, Judy A. *NASA Historical Data Book, Vol. VI: NASA Space Applications, Aeronautics and Space Research and Technology, Tracking and Data Acquisition/Support Operations, Commercial Programs, and Resources, 1979-1988.* NASA SP-4012, 1999.

Noordung, Hermann. *The Problem of Space Travel: The Rocket Motor.* Edited by Ernst Stuhlinger and J.D. Hunley, with Jennifer Garland. NASA SP-4026, 1995.

Gawdiak, Ihor Y., Ramon J. Miro, and Sam Stueland, comps. *Astronautics and Aeronautics, 1986-1990: A Chronology.* NASA SP-4027, 1997.

Gawdiak, Ihor Y. and Shetland, Charles. *Astronautics and Aeronautics, 1991-1995: A Chronology.* NASA SP-2000-4028, 2000.

Orloff, Richard W. *Apollo by the Numbers: A Statistical Reference.* NASA SP-2000-4029, 2000. Visit http://history.nasa.gov/series95.html for a pdf version of this document..

Management Histories, NASA SP-4100

Rosholt, Robert L. *An Administrative History of NASA, 1958-1963.* NASA SP-4101, 1966.

Levine, Arnold S. *Managing NASA in the Apollo Era.* NASA SP-4102, 1982.

Roland, Alex. *Model Research: The National Advisory Committee for Aeronautics, 1915-1958.* NASA SP-4103, 1985.

Fries, Sylvia D. NASA *Engineers and the Age of Apollo.* NASA SP-4104, 1992.

Glennan, T. Keith. *The Birth of NASA: The Diary of T. Keith Glennan.* Edited by J.D. Hunley. NASA SP-4105, 1993.

Seamans, Robert C. *Aiming at Targets: The Autobiography of Robert C. Seamans.* NASA SP-4106, 1996.

Garber, Stephen J., editor. *Looking Backward, Looking Forward: Forty Years of Human Spaceflight Symposium.* NASA SP-2002-4107.

Mallick, Donald L. with Peter W. Merlin. *The Smell of Kerosene: A Test Pilot's Odyssey.* NASA SP-4108.

Iliff, Kenneth W. and Curtis L. Peebles. *From Runway to Orbit: Reflections of a NASA Engineer.* NASA SP-2004-4109.

Laufer, Alexander, Post, Todd, and Hoffman, Edward. *Shared Voyage: Learning and Unlearning from Remarkable Projects.* NASA SP-2005-4111.

Dawson, Virginia P. and Bowles, Mark D. *Realizing the Dream of Flight: Biographical Essays in Honor of the Centennial of Flight, 1903-2003.* NASA SP-2005-4112.

Project Histories, NASA SP-4200:

Swenson, Loyd S., Jr., James M. Grimwood, and Charles C. Alexander. *This New Ocean: A History of Project Mercury.* NASA SP-4201, 1966, reprinted 1999.

Green, Constance McLaughlin, and Milton Lomask. *Vanguard: A History.* NASA SP-4202, 1970; rep. ed. Smithsonian Institution Press, 1971.

Hacker, Barton C., and James M. Grimwood. *On Shoulders of Titans: A History of Project Gemini.* NASA SP-4203, 1977, reprinted 2002.

Benson, Charles D. and William Barnaby Faherty. *Moonport: A History of Apollo Launch Facilities and Operations.* NASA SP-4204, 1978. University Press of Florida has republished the book in two volumes, *Gateway to the Moon* and *Moon Launch!*

Brooks, Courtney G., James M. Grimwood, and Loyd S. Swenson, Jr. *Chariots for Apollo: A History of Manned Lunar Spacecraft.* NASA SP- 4205, 1979.

Bilstein, Roger E. *Stages to Saturn: A Technological History of the Apollo/Saturn Launch Vehicles.* NASA SP-4206, 1980 and 1996. Reprinted by the University Press of Florida.

Compton, W. David, and Charles D. Benson. *Living and Working in Space: A History of Skylab.* NASA SP-4208, 1983.

Ezell, Edward Clinton, and Linda Neuman Ezell. *The Partnership: A History of the Apollo-Soyuz Test Project.* NASA SP-4209, 1978.

Hall, R. Cargill. *Lunar Impact: A History of Project Ranger.* NASA SP-4210, 1977.

Newell, Homer E. *Beyond the Atmosphere: Early Years of Space Science.* NASA SP-4211, 1980.

Ezell, Edward Clinton, and Linda Neuman Ezell. *On Mars: Exploration of the Red Planet, 1958-1978.* NASA SP-4212, 1984.

Pitts, John A. *The Human Factor: Biomedicine in the Manned Space Program to 1980.* NASA SP-4213, 1985.

Compton, W. David. *Where No Man Has Gone Before: A History of Apollo Lunar Exploration Missions.* NASA SP-4214, 1989.

Naugle, John E. *First Among Equals: The Selection of NASA Space Science Experiments.* NASA SP-4215, 1991.

Wallace, Lane E. *Airborne Trailblazer: Two Decades with NASA Langley's 737 Flying Laboratory.* NASA SP-4216, 1994.

Butrica, Andrew J. *Beyond the Ionosphere: Fifty Years of Satellite Communications.* NASA SP-4217, 1997.

Butrica, Andrew J. *To See the Unseen: A History of Planetary Radar Astronomy.* NASA SP-4218, 1996.

Mack, Pamela E., ed. *From Engineering Science to Big Science: The NACA and NASA Collier Trophy Research Project Winners.* NASA SP-4219, 1998.

Reed, R. Dale. *Wingless Flight: The Lifting Body Story.* NASA SP-4220, 1998.

Heppenheimer, T. A. *The Space Shuttle Decision: NASA's Search for a Reusable Space Vehicle.* NASA SP-4221, 1999.

Hunley, J. D., ed. *Toward Mach 2: The Douglas D-558 Program.* NASA SP-4222, 1999.

Swanson, Glen E., ed. *"Before This Decade is Out..." Personal Reflections on the Apollo Program.* NASA SP-4223, 1999.

Tomayko, James E. *Computers Take Flight: A History of NASA's Pioneering Digital Fly-By-Wire Project.* NASA SP-4224, 2000.

Morgan, Clay. *Shuttle-Mir: The United States and Russia Share History's Highest Stage.* NASA SP-2001-4225.

Leary, William M. *We Freeze to Please: A History of NASA's Icing Research Tunnel and the Quest for Safety.* NASA SP-2002-4226, 2002.

Mudgway, Douglas J. *Uplink-Downlink: A History of the Deep Space Network, 1957-1997.* NASA SP-2001-4227.

Dawson, Virginia P. and Mark D. Bowles. *Taming Liquid Hydrogen: The Centaur Upper Stage Rocket, 1958-2002.* NASA SP-2004-4230.

Meltzer, Michael. *Mission to Jupiter: A History of the Galileo Project.* NASA SP-2007-4231.

Heppenheimer, T.A. *Facing the Heat Barrier: A History of Hypersonics.* NASA SP-2007-4232. Visit http://history.nasa.gov/series 95.html for a pdf version of this document.

Tsiao, Sunny. *"Read You Loud and Clear!" The Story of NASA's Spaceflight Tracking and Data Network.* NASA SP-2007-4233.

Center Histories, NASA SP-4300:

Rosenthal, Alfred. *Venture into Space: Early Years of Goddard Space Flight Center.* NASA SP-4301, 1985.

Hartman, Edwin, P. *Adventures in Research: A History of Ames Research Center, 1940-1965.* NASA SP-4302, 1970.

Hallion, Richard P. *On the Frontier: Flight Research at Dryden, 1946-1981.* NASA SP-4303, 1984.

Muenger, Elizabeth A. *Searching the Horizon: A History of Ames Research Center, 1940-1976.* NASA SP-4304, 1985.

Hansen, James R. *Engineer in Charge: A History of the Langley Aeronautical Laboratory, 1917-1958.* NASA SP-4305, 1987.

Dawson, Virginia P. *Engines and Innovation: Lewis Laboratory and American Propulsion Technology.* NASA SP-4306, 1991.

Dethloff, Henry C. *"Suddenly Tomorrow Came...": A History of the Johnson Space Center, 1957-1990.* NASA SP-4307, 1993.

Hansen, James R. *Spaceflight Revolution: NASA Langley Research Center from Sputnik to Apollo.* NASA SP-4308, 1995.

Wallace, Lane E. *Flights of Discovery: An Illustrated History of the Dryden Flight Research Center.* NASA SP-4309, 1996.

Herring, Mack R. *Way Station to Space: A History of the John C. Stennis Space Center.* NASA SP-4310, 1997.

Wallace, Harold D., Jr. *Wallops Station and the Creation of an American Space Program.* NASA SP-4311, 1997.

Wallace, Lane E. *Dreams, Hopes, Realities. NASA's Goddard Space Flight Center: The First Forty Years.* NASA SP-4312, 1999.

Dunar, Andrew J. and Waring, Stephen P. *Power to Explore: A History of Marshall Space Flight Center, 1960-1990.* NASA SP-4313, 1999.

Bugos, Glenn E. *Atmosphere of Freedom: Sixty Years at the NASA Ames Research Center.* NASA SP-2000-4314, 2000.

Schultz, James. *Crafting Flight: Aircraft Pioneers and the Contributions of the Men and Women of NASA Langley Research Center.* NASA SP-2003-4316, 2003.

Bowles, Mark D. *Science in Flux: NASA's Nuclear Program at Plum Brook Station, 1955-2005.* NASA SP-2006-4317.

Wallace, Lane E. *Flights of Discovery: Sixty Years of Flight Research at the Dryden Flight Research Center.* NASA SP-4318, 2007.

General Histories, NASA SP-4400:

Corliss, William R. NASA *Sounding Rockets, 1958-1968: A Historical Summary.* NASA SP-4401, 1971.

Wells, Helen T., Susan H. Whiteley, and Carrie Karegeannes. *Origins of NASA Names.* NASA SP-4402, 1976.

Anderson, Frank W., Jr. Orders of Magnitude: A History of NACA and NASA, 1915-1980. NASA SP-4403, 1981.

Sloop, John L. *Liquid Hydrogen as a Propulsion Fuel, 1945-1959.* NASA SP-4404, 1978.

Roland, Alex. *A Spacefaring People: Perspectives on Early Spaceflight.* NASA SP-4405, 1985.

Bilstein, Roger E. *Orders of Magnitude: A History of the NACA and NASA, 1915-1990.* NASA SP-4406, 1989.

Siddiqi, Asif A., *Challenge to Apollo: The Soviet Union and the Space Race, 1945-1974.* NASA SP-2000-4408, 2000.

Hogan, Thor. *Mars Wars: The Rise and Fall of the Space Exploration Initiative.* NASA SP-2007-4410, 2007.

Monographs in Aerospace History (SP-4500 Series):

Launius, Roger D. and Aaron K. Gillette, comps. *Toward a History of the Space Shuttle: An Annotated Bibliography.* Monograph in Aerospace History, No. 1, 1992.

Launius, Roger D., and J.D. Hunley, comps. *An Annotated Bibliography of the Apollo Program.* Monograph in Aerospace History No. 2, 1994.

Launius, Roger D. *Apollo: A Retrospective Analysis.* Monograph in Aerospace History, No. 3, 1994.

Hansen, James R. *Enchanted Rendezvous: John C. Houbolt and the Genesis of the Lunar-Orbit Rendezvous Concept.* Monograph in Aerospace History, No. 4, 1995.

Gorn, Michael H. *Hugh L. Dryden's Career in Aviation and Space.* Monograph in Aerospace History, No. 5, 1996.

Powers, Sheryll Goecke. *Women in Flight Research at NASA Dryden Flight Research Center from 1946 to 1995.* Monograph in Aerospace History, No. 6, 1997.

Portree, David S.F. and Robert C. Trevino. *Walking to Olympus: An EVA Chronology.* Monograph in Aerospace History, No. 7, 1997.

Logsdon, John M., moderator. *Legislative Origins of the National Aeronautics and Space Act of 1958: Proceedings of an Oral History Workshop.* Monograph in Aerospace History, No. 8, 1998. Visit http://history.nasa.gov/series95.html for a pdf version of this document.

Rumerman, Judy A., comp. *U.S. Human Spaceflight, A Record of Achievement 1961-1998.* Monograph in Aerospace History, No. 9, 1998.

Portree, David S. F. *NASA's Origins and the Dawn of the Space Age.* Monograph in Aerospace History, No. 10, 1998.

Logsdon, John M. *Together in Orbit: The Origins of International Cooperation in the Space Station.* Monograph in Aerospace History, No. 11, 1998.

Phillips, W. Hewitt. *Journey in Aeronautical Research: A Career at NASA Langley Research Center.* Monograph in Aerospace History, No. 12, 1998.

Braslow, Albert L. *A History of Suction-Type Laminar-Flow Control with Emphasis on Flight Research.* Monograph in Aerospace History, No. 13, 1999.

Logsdon, John M., moderator. *Managing the Moon Program: Lessons Learned From Apollo.* Monograph in Aerospace History, No. 14, 1999.

Perminov, V.G. *The Difficult Road to Mars: A Brief History of Mars Exploration in the Soviet Union.* Monograph in Aerospace History, No. 15, 1999. Visit http://history.nasa.gov/series95.html for a pdf version of this document..

Tucker, Tom. *Touchdown: The Development of Propulsion Controlled Aircraft at NASA Dryden.* Monograph in Aerospace History, No. 16, 1999.

Maisel, Martin, Giulanetti, Demo J., and Dugan, Daniel C. *The History of the XV-15 Tilt Rotor Research Aircraft: From Concept to Flight.* Monograph in Aerospace History, No. 17, 2000.

Jenkins, Dennis R. *Hypersonics Before the Shuttle: A Concise History of the X-15 Research Airplane.* Monograph in Aerospace History, No. 18, 2000.

Chambers, Joseph R. *Partners in Freedom: Contributions of the Langley Research Center to U.S. Military Aircraft of the 1990s.* Monograph in Aerospace History, No. 19, 2000 (NASA SP-2000-4519).

Waltman, Gene L. *Black Magic and Gremlins: Analog Flight Simulations at NASA's Flight Research Center.* Monograph in Aerospace History, No. 20, 2000 (NASA SP-2000-4520).

Portree, David S.F. *Humans to Mars: Fifty Years of Mission Planning, 1950-2000.* Monograph in Aerospace History, No. 21, 2001 (NASA SP-2001-4521).

Thompson, Milton O. with J.D. Hunley. *Flight Research: Problems Encountered and What They Should Teach Us.* Monograph in Aerospace History, No. 22, 2001 (NASA SP-2001-4522).

Tucker, Tom. *The Eclipse Project.* Monograph in Aerospace History, No. 23, 2001 (NASA SP-2001-4523).

Siddiqi, Asif A. *Deep Space Chronicle: A Chronology of Deep Space and Planetary Probes 1958-2000.* Monograph in Aerospace History, No. 24, 2002 (NASA SP-2002-4524).

Merlin, Peter W. *Mach 3+: NASA/USAF YF-12 Flight Research, 1969-1979.* Monograph in Aerospace History, No. 25, 2001 (NASA SP-2001-4525).

Anderson, Seth B. *Memoirs of an Aeronautical Engineer: Flight Tests at Ames Research Center: 1940-1970.* Monograph in Aerospace History, No. 26, 2002 (NASA SP-2002-4526)

Renstrom, Arthur G. *Wilbur and Orville Wright: A Bibliography Commemorating the One-Hundredth Anniversary of the First Powered Flight on December 17, 1903.* Monograph in Aerospace History, No. 27, 2002 (NASA SP-2002-4527).

No monograph 28.

Chambers, Joseph R. *Concept to Reality: Contributions of the NASA Langley Research Center to U.S. Civil Aircraft of the 1990s.* Monograph in Aerospace History, No. 29, 2003. (SP-2003-4529).

Peebles, Curtis, editor. *The Spoken Word: Recollections of Dryden History, The Early Years.* Monograph in Aerospace History, No. 30, 2003. (SP-2003-4530).

Jenkins, Dennis R., Tony Landis, and Jay Miller. *American X-Vehicles: An Inventory- X-1 to X-50.* Monograph in Aerospace History, No. 31, 2003 (SP-2003-4531).

Renstrom, Arthur G. *Wilbur and Orville Wright: A Chronology Commemorating the One-Hundredth Anniversary of the First Powered Flight on December 17, 1903.* Monograph in Aerospace History, No. 32, 2003. (NASA SP-2003-4532).

Bowles, Mark D. and Arrighi, Robert S. *NASA's Nuclear Frontier: The Plum Brook Research Reactor.* Monograph in Aerospace History, No. 33, 2004. (SP-2004-4533).

Matranga, Gene J.; Wayne C. Ottinger; Calvin R.; Jarvis, and D. Christian Gelzer. *Unconventional, Contrary, and Ugly: The Lunar Landing Research Vehicle.* Monograph in Aerospace History, No. 35, 2006. (NASA SP-2004-4535).

McCurdy, Howard E. *Low Cost Innovation in Spaceflight: The History of the Near Earth Asteroid Rendezvous (NEAR) Mission.* Monograph in Aerospace History, No. 36, 2005. (NASA SP-2005-4536).

Seamans, Robert C. Jr. *Project Apollo: The Tough Decisions.* Monograph in Aerospace History, No. 37, 2005. (NASA SP-2005-4537).

Lambright, W. Henry. *NASA and the Environment: The Case of Ozone Depletion.* Monograph in Aerospace History, No. 38, 2005. (NASA SP-2005-4538).

Chambers, Joseph R. *Innovation in Flight: Research of the NASA Langley Research Center on Revolutionary Advanced Concepts for Aeronautics.* Monograph in Aerospace History, No. 39, 2005. (NASA SP-2005-4539). This monograph is only available on-line. Visit http://history.nasa.gov/series95.html.

Phillips, W. Hewitt. *Journey Into Space Research: Continuation of a Career at NASA Langley Research Center.* Monograph in Aerospace History, No. 40, 2005. (NASA SP-2005-4540). This monograph is only available on-line. Visit http://history.nasa.gov/series95.html.

Rumerman, Judy A., comp. *U.S. Human Spaceflight: A Record of Achievement, 1961-2006.* Monograph in Aerospace History No. 41, 2007. (NASA SP-2007-4541). This is an updating by Chris Gamble and Gabriel Okolski of the similarly titled Monograph 9 that was published in 1998.

Dryden Historical Studies

Tomayko, James E., and Christian Gelzer, editor. *The Story of Self-Repairing Flight Control Systems.* Dryden Historical Study #1.

Dana, William H. *X-38: Flight Testing the Prototype Crew Recovery Vehicle.* Dryden Historical Study #2.

Electronic Media (SP-4600 Series)

Remembering Apollo 11: The 30th Anniversary Data Archive CD-ROM. (NASA SP-4601, 1999)

Remembering Apollo 11: The 35th Anniversary Data Archive CD-ROM. (NASA SP-2004-4601, 2004). This is an update of the 1999 edition.

The Mission Transcript Collection: U.S. Human Spaceflight Missions from Mercury Redstone 3 to Apollo 17. (SP-2000-4602, 2001). Available commercially from CG Publishing.

Shuttle-Mir: the United States and Russia Share History's Highest Stage. (NASA SP-2001-4603, 2002). This CD-ROM is available from NASA CORE.

U.S. Centennial of Flight Commission presents Born of Dreams ~ Inspired by Freedom. (NASA SP-2004-4604, 2004).

Of Ashes and Atoms: A Documentary on the NASA Plum Brook Reactor Facility. (NASA SP-2005-4605).

Taming Liquid Hydrogen: The Centaur Upper Stage Rocket Interactive CD-ROM. (NASA SP-2004-4606, 2004).

Fueling Space Exploration: The History of NASA's Rocket Engine Test Facility DVD. (NASA SP-2005-4607).

Altitude Wind Tunnel at NASA Glenn Research Center: An Interactive History. (NASA SP-2008-4608).

Conference Proceedings (SP-4700 Series)

Dick, Steven J. and Cowing, Keith L, ed. *Risk and Exploration: Earth, Sea and the Stars.* (NASA SP-2005-4701).

Dick, Steven J. and Launius, Roger D. *Critical Issues in the History of Spaceflight.* (NASA SP-2006-4702).

Historical Reports (NASA HHR)

Boone, W. Fred. *NASA Office of Defense Affairs: The First Five Years.* (NASA HHR-32, 1970).

Research in NASA History: A Guide to the NASA History Program. (NASA HHR-64, revised June 1997).

NASA Special Reports (NASA SP-4900)

Kloman, Erasmus H. *Unmanned Space Project Management: Surveyor and Lunar Orbiter.* (NASA SP-4901, 1972).

Other NASA Special Publications, not in the formal NASA History Series

The Impact of Science on Society. NASA SP-482 by James Burke, Jules Bergman, and Isaac Asimov, 1985.

Space Station Requirements and Transportation Options for Lunar Outpost. NASA, 1990.

Space Station Freedom Accommodation of the Human Exploration Initiative. NASA, 1990.

Why Man Explores. NASA EP-125, 1976.

Results of the Second Manned Suborbital Space Flight, July 21, 1961. NASA, 1961.

Results of the Second U.S. Manned Orbital Space Flight. NASA SP-6, 1962.

Results of the Third U.S. Manned Orbital Space Flight. NASA SP-12, 1962.

Mercury Project Summary including Results of the Fourth Manned Orbital Flight. NASA SP-45, 1963.

X-15 Research Results With a Selected Bibliography. NASA SP-60, 1965.

Exploring Space with a Camera. NASA SP-168, 1968.

Aerospace Food Technology. NASA SP-202, 1969.

What Made Apollo a Success? NASA SP-287, 1971.

Evolution of the Solar System. NASA SP-345, 1976.

Pioneer Odyssey (NASA SP-349/396, revised edition, 1977) by Richard Fimmel, William Swindell, and Eric Burgess.

Apollo Expeditions to the Moon. NASA SP-350, 1975.

Apollo Over the Moon: A View From Orbit. (NASA SP-362, 1978) edited by Harold Masursky, G.W. Colton, and Farouk El-Baz.

Introduction to the Aerodynamics of Flight. (NASA SP-367, 1975) by Theodore A. Talay.

Biomedical Results of Apollo. (NASA SP-368, 1975), edited by Richard S. Johnston, Lawrence F. Dietlein, M.D., and Charles A. Berry, M.D.

Skylab EREP Investigations Summary. (NASA SP-399, 1978)

Skylab: Our First Space Station. (NASA SP-400, 1977), edited by Leland F. Belew.

Skylab, Classroom in Space. (NASA SP-401, 1977), edited by Lee Summerlin.

A New Sun: Solar Results from Skylab. (NASA SP-402, 1979) by John A. Eddy and edited by Rein Ise.

Skylab's Astronomy and Space Sciences. (NASA SP-404, 1979), edited by Charles A. Lundquist.

The Space Shuttle. (NASA SP-407, 1976)

The Search For Extraterrestrial Intelligence. (NASA SP-419, 1977), edited by Philip Morrison, John Billingham, and John Wolfe.

Atlas of Mercury. (NASA SP-423, 1978) by Merton E. Davies, Stephen E. Dwornik, et. al.

The Voyage of Mariner 10: Mission to Venus and Mercury. (NASA SP-424, 1978) by James A. Dunne and Eric Burgess.

The Martian Landscape. (NASA SP-425, 1978)

The Space Shuttle at Work. (NASA SP-432/EP-156 1979) by Howard Allaway.

Project Orion: A Design Study of a System for Detecting Extrasolar Planets. (NASA SP-436, 1980), edited by David C. Black.

Wind Tunnels of NASA. (NASA SP-440, 1981).

Viking Orbiter Views of Mars. (NASA SP-441, 1980)

The High Speed Frontier: Case Histories of Four NACA Programs, 1920-1950. (NASA SP-445, 1980.)

The Star Splitters: The High Energy Astronomy Observatories. (NASA SP-466, 1984) by Wallace H. Tucker.

Planetary Geology in the 1980s. (NASA SP-467, 1985) by Joseph Veverka.

Quest for Performance: The Evolution of Modern Aircraft. (NASA SP-468, 1985.)

The Long Duration Exposure Facility (LDEF): Mission 1 Experiments. (SP-473, 1984) ed. by Lenwood G. Clark, William H. Kinar, et. al.

Voyager 1 and 2, Atlas of Saturnian Satellites. (NASA SP-474, 1984) edited by Raymond Batson.

Far Travelers: The Exploring Machines. (NASA SP-480, 1985) by Oran W. Nicks.

Living Aloft: Human Requirements for Extended Spaceflight. (NASA SP-483, 1985)

Space Shuttle Avionics System. (NASA SP-504, 1989) by John F. Hanaway and Robert W. Moorehead.

Life Into Space: Space Life Sciences Research, Volumes I - III. 1965-2003 (NASA SP-534).

Flight Research at Ames, 1940-1997. (NASA SP-3300, 1998).

The Planetary Quarantine Program. (NASA SP-4902, 1974).

Spaceborne Digital Computer Systems. (NASA SP-8070, 1971).

Magellan: The Unveiling of Venus. (JPL-400-345, 1989)

Guide to Magellan Image Interpretation. (JPL-93-24) by John Ford, Jeffrey Plaut, et. al.

The Apollo Program Summary Report. (Document # JSC-09423, April 1975)

Saturn Illustrated Chronology. (MHR-5, Marshall Space Flight Center, fifth edition, 1971) prepared by David S. Akens.

Celebrating a Century of Flight. (NASA SP-2002-09-511-HQ). Edited by Tony Springer.

Present and Future State of the Art in Guidance Computer Memories. (NASA TN D-4224, 1967) by Robert C. Ricci.

NASA Educational Publications

Apollo 13 "Houston, we've got a problem." (NASA EP-76, 1970).

On the Moon with Apollo 16: A Guide to the Descartes Region. (NASA EP-95, 1972)

Skylab: A Guidebook. (NASA EP-107, 1973), by Leland F. Belew and Ernst Stuhlinger.

Spacelab: An International Short-Stay Orbiting Laboratory. (NASA EP-165) by Walter Froehlich.

A Meeting with the Universe: Science Discoveries from the Space Program. (NASA EP-177, 1981).

NASA Publications (NPs)

Science in Orbit: The Shuttle & Spacelab Experience: 1981-1986. (NASA NP-119, 1988).

NASA Conference Proceedings

Life in the Universe: Proceedings of a conference held at NASA Ames Research Center Moffet Field, California, June 19-20, 1979. (NASA CP-2156, 1981), edited by John Billingham.

Proceedings of the X-15 First Flight 30th Anniversary Celebration of June 8, 1989.

NASA Technical Memoranda

Destination Moon: A History of the Lunar Orbiter Program. (NASA TM-3487, 1977) by Bruce Byers.

New Series in NASA History Published by the American Institute of Aeronautics and Astronautics.

Peebles, Curtis. *The X-43A Flight Research Program: Lessons Learned on the Road to Mach 10.* AIAA, 2008.

Merlin, Peter. *From Archangel to Senior Crown: The Design and Development of the Blackbird.* AIAA, 2008.

New Series in NASA History Published by the Johns Hopkins University Press:

Cooper, Henry S. F., Jr. *Before Lift-off: The Making of a Space Shuttle Crew.* Baltimore: Johns Hopkins University Press, 1987.

McCurdy, Howard E. *The Space Station Decision: Incremental Politics and Technological Choice.* Baltimore: Johns Hopkins University Press, 1990.

Hufbauer, Karl. *Exploring the Sun: Solar Science Since Galileo.* Baltimore: Johns Hopkins University Press, 1991.

McCurdy, Howard E. *Inside NASA: High Technology and Organizational Change in the U.S. Space Program.* Baltimore: Johns Hopkins University Press, 1993.

Lambright, W. Henry. *Powering Apollo: James E. Webb of NASA.* Baltimore: Johns Hopkins University Press, 1995.

Bromberg, Joan Lisa. *NASA and the Space Industry.* Baltimore: Johns Hopkins University Press, 1999.

Beattie, Donald A. *Taking Science to the Moon: Lunar Experiments and the Apollo Program.* Baltimore: Johns Hopkins University Press, 2001.

McCurdy, Howard E. Faster, *Better, Cheaper: Low-Cost Innovation in the U.S. Space Program.* Baltimore: Johns Hopkins University Press, 2001.

Johnson, Stephen B. *The Secret of Apollo: Systems Management in American and European Space Programs.* Baltimore: Johns Hopkins University Press, 2002.

Lambright, W. Henry, editor. *Space Policy in the 21st Century.* Baltimore: Johns Hopkins University Press, 2002.

Bilstein, Roger E. *Testing Aircraft, Exploring Space: An Illustrated History of NACA and NASA.* Baltimore: Johns Hopkins University Press, 2003.

Butrica, Andrew J. *Single Stage to Orbit: Politics, Space Technology, and the Quest for Reusable Rocketry.* Baltimore: Johns Hopkins University Press, 2005.

Conway, Erik M. *High-Speed Dreams: NASA and the Technopolitics of Supersonic Transportation, 1945-1999.* Baltimore: Johns Hopkins University Press, 2005.

Launius, Roger D. and Howard E. McCurdy. *Robots in Space: Technology, Evolution, and Interplanetary Travel.* Baltimore: Johns Hopkins University Press, 2008.

Conway, Erik M. *Atmospheric Science at NASA: A History.* Baltimore: Johns Hopkins University Press, 2008.

NASA History Titles Published by Texas A&M University Press

Schorn, Ronald A. *Planetary Astronomy: From Ancient Times to the Third Millennium*. College Station: Texas A&M University Press, 1998.

NASA History Titles Published by The University Press of Kentucky

Gorn, Michael H. *Expanding the Envelope: Flight Research at NACA and NASA*. Lexington: The University Press of Kentucky, 2001.

Reed, R. Dale. *Wingless Flight: The Lifting Body Story*. Lexington: The University Press of Kentucky, 2002.

Ed. by Launius, Roger D. and Dennis R. Jenkins. *To Reach the High Frontier: A History of U.S. Launch Vehicles*. Lexington: The University Press of Kentucky, 2002.

NASA History Titles Published by the University Press of Florida

Ed. by Swanson, Glen W. *"Before This Decade is Out...": Personal Reflections on the Apollo Program*. Gainesville: The University Press of Florida, 2002.

Benson, Charles D. and William B. Faherty. *Moon Launch!: A History of the Saturn-Apollo Launch Operations*. Gainesville: The University Press of Florida, 2001.

Benson, Charles D. and William B. Faherty. *Gateway to the Moon: Building the Kennedy Space Center Launch Complex*. Gainesville: The University Press of Florida, 2001.

Bilstein, Roger E. *Stages to Saturn: A Technological History of the Apollo/Saturn Launch Vehicles*. (Originally published as NASA SP-4206 in 1980 and reprinted in 1996). Gainesville: The University Press of Florida, 2003.

Siddiqi, Asif A. *The Soviet Space Race with Apollo*. Gainesville: The University Press of Florida, 2003.

Siddiqi, Asif A. *Sputnik and the Soviet Space Challenge*. Gainesville: The University Press of Florida, 2003.

Lipartito, Kenneth and Butler, Orville R. *A History of the Kennedy Space Center*. Gainesville: The University Press of Florida, 2007.

NASA History Titles Published by Harwood Academic Press

Ed. by Roger D. Launius, John M. Logsdon and Robert W. Smith. *Reconsidering Sputnik: Forty Years Since the Soviet Satellite*. London: Harwood Academic Press, 2000.

NASA History Titles Published by the University of Illinois Press

Ed. by Roger D. Launius and Howard McCurdy. *Spaceflight and the Myth of Presidential Leadership*. Urbana, IL: University of Illinois Press, 1997.

NASA History Titles Published by Greenwood Press

Launius, Roger D. *Frontiers of Space Exploration.* Westport, CT: Greenwood Press, 1998.

NASA History Titles Published by the Smithsonian Institution Press

Heppenheimer, T.A. *Development of the Shuttle, 1972-1981.* Washington, DC: Smithsonian Institution Press, 2002.

Dethloff, Henry C. and Ronald A. Schorn. *Voyager's Grand Tour: To the Outer Planets and Beyond.* Washington, DC: Smithsonian Institution Press, 2003.

Hallion, Richard P. and Michael H. Gorn. *On the Frontier: Experimental Flight at NASA Dryden.* Washington, DC: Smithsonian Institution Press, 2003.

NASA History Titles Published by CG Publishing, Inc.

The Mission Transcript Collection: U.S. Human Spaceflight Missions From Mercury Redstone 3 to Apollo 17. (NASA SP-2000-4602).

NASA History Titles Published by Abrams Press

Dick, Steven, editor, et. al. *America In Space: NASA's First Fifty Years.* New York: Abrams, 2007.

America in Space, published by Harry N. Abrams, Inc., New York.

Miscellaneous Publications of NASA History

Dawson, Virginia. *Ideas Into Hardware: A History of the Rocket Engine Test Facility at the NASA Glenn Research Center.* Cleveland, 2004.